CW00908543

X.media.publishing

Wolfgang Kühn · Martin Grell

JDF

Process Integration, Technology, Product Description

With 20 Figures and 8 Tables

Prof. Dr.-Ing. Wolfgang Kühn
University of Wuppertal
Faculty of Electrical, Information and Media Engineering
Rainer-Gruenter-Str. 21
42119 Wuppertal
Germany
wkuehn@uni-wuppertal.de

Martin Grell
Paulinenplatz 5
20359 Hamburg
Germany
martin.grell@web.de

Translated from the German „JDF" (Springer-Verlag 2004, ISBN 3-540-20758-9) by Derek Robinson, Linguatext, Edinburgh, Scotland; www.linguatext.com.

Library of Congress Control Number: 2005921089

ISSN 1612-1449
ISBN 3-540-23560-4 Springer Berlin Heidelberg New York

Springer is a part of Springer Science+Business Media
springeronline.com

Printed in The Netherlands

Cover design: KünkelLopka, Heidelberg
Typesetting and Production: LE-TeX Jelonek, Schmidt & Vöckler GbR, Leipzig
Printed on acid-free paper 33/3142/YL - 5 4 3 2 1 0

Table of Contents

1 Introduction 1

2 Process integration in the print media industry 3

2.1 The essential problem with non-networked production – process costs 4
2.2 Data types in the print media industry 5
2.2.1 Content data 6
2.2.2 Master data 6
2.2.3 Job data 6
2.2.4 Production data 7
2.2.5 Control data 8
2.2.6 Operating and machine data 9
2.2.7 Quality data 9
2.3 Networking routes of process integration 9
2.3.1 E-business 11
2.3.2 Job preparation 12
2.3.3 Machine presetting 13
2.3.4 Production planning and control 15
2.3.5 Operating data logging and actual costing 16
2.3.6 Color workflow 17
2.4 Why has process integration failed so often in the past? 18

3 Job Definition Format **21**

3.1 Who needs to understand what about JDF? 23
3.2 JDF basics 23
3.2.1 Interfaces 24
3.2.2 How does JDF work? 26
3.2.3 Job Messaging Format (JMF) 29
3.2.4 Private sections 30
3.3 Implementation of JDF by CIP4 30

3.4 Interrelation with other print media industry standards 31
3.4.1 JDF and PPF, PJTF, IFRAtrack 31
3.4.2 JDF and Print Talk 32
3.4.3 JDF and PDF/X3 32
3.4.4 JDF and PPML/VDX 33
3.4.5 JDF and EDIFACT 33
3.5 How secure is JDF? 34

4 Networking architectures **35**
4.1 Networking architecture basics 36
4.2 Prinect from Heidelberger Druckmaschinen 37
4.2.1 Prinect layer: Applications 39
4.2.2 Prinect layer: Centralized services 40
4.2.3 Prinect layer: JDF processors 40
4.3 PrintCity "Closed Loop... Open Systems" 42
4.4 Networked Graphic Production (NGP) 43
4.5 Order management system – a JDF nerve center 45
4.6 EFI 45
4.7 PrintNet (ppi Media) 46
4.8 Evaluation 47

5 Benefits **49**
5.1 Benefits of e-business 49
5.2 The benefits of networked job preparation 52
5.3 The benefits of networked machine presetting 53
5.4 The benefits of networked production planning and control 54
5.5 The benefits of networked operating data logging and actual costing 55
5.6 The benefits of the networked color workflow 57
5.7 Payback period of a networking project 58
5.7.1 Basis for calculating the payback period 59
5.7.2 Practical example 60
5.7.3 Quantitative benefits of networking 61
5.7.4 Evaluation of the investment decision 62

6 Procedure when implementing process integration **65**
6.1 Prerequisites for networking 65
6.1.1 Technical prerequisites 65
6.1.2 Organizational prerequisites 66
6.2 Step-by-step implementation 67
6.2.1 Requirements analysis 67

6.2.2 Process costs analysis 68
6.2.3 Choosing a suitable partner.......................... 70
6.2.4 Organizational measures.............................. 71
6.2.5 Technical implementation of networking......... 73
6.3 Verifying success and ongoing optimization............. 74
6.3.1 Duties of the controlling department 74
6.3.2 Key management statistics 75

7 Closing remarks **77**

Annex **79**

Checklist for reviewing process inefficiencies 79
Checklist for devising a networking concept.................... 81
Organisations 84
Companies 86
Industry standards 88
Glossary... 90

Subject Index **97**

1 Introduction

... JDF may not be as headline-grabbing as a new press but it's potentially the most important development in the printing industry since postscript.

Simon Eccles, Electronic Imaging Magazine

The print media industry is facing tough competition. Overcapacity and sluggish demand are cutting margins. At the same time, customers' expectations as regards flexibility, quality, speed and reliability have never been higher. This situation calls for processes to be reviewed and manufacturing costs to be cut even further.

The various machines and processes have already been widely optimized in the past. The next challenge facing the print media industry is to resolve the problem of software "pockets" that exist in isolation. Process integration is the goal, and the technology to achieve this is provided by the newly-developed Job Definition Format (JDF).

Supported by the PDF document format and vendor-independent JDF, a new generation of fully integrated workflow solutions is currently being developed that enable process integration along the entire value added chain. Print service providers who network their systems have opportunities to cut process costs, but also lay themselves open to the risks associated with deploying new technologies. In order to safeguard their investment, therefore, companies are seeing an increasing need to source comprehensive technical and business background information.

This book is aimed at everyone who is facing decisions on process integration and networking in the print media industry or who is involved in such decisions either directly or indirectly. Readers are provided with focussed information supporting investment decisions and successful implementation of networking projects.

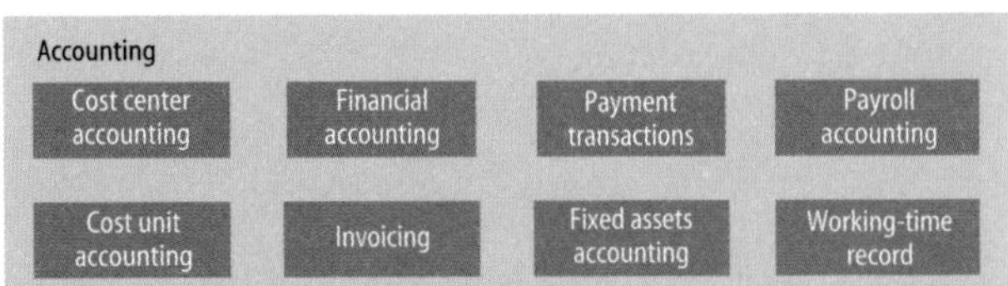
Accounting
Cost center accounting
Financial accounting
Payment transactions
Payroll accounting
Cost unit accounting
Invoicing
Fixed assets accounting
Working-time record

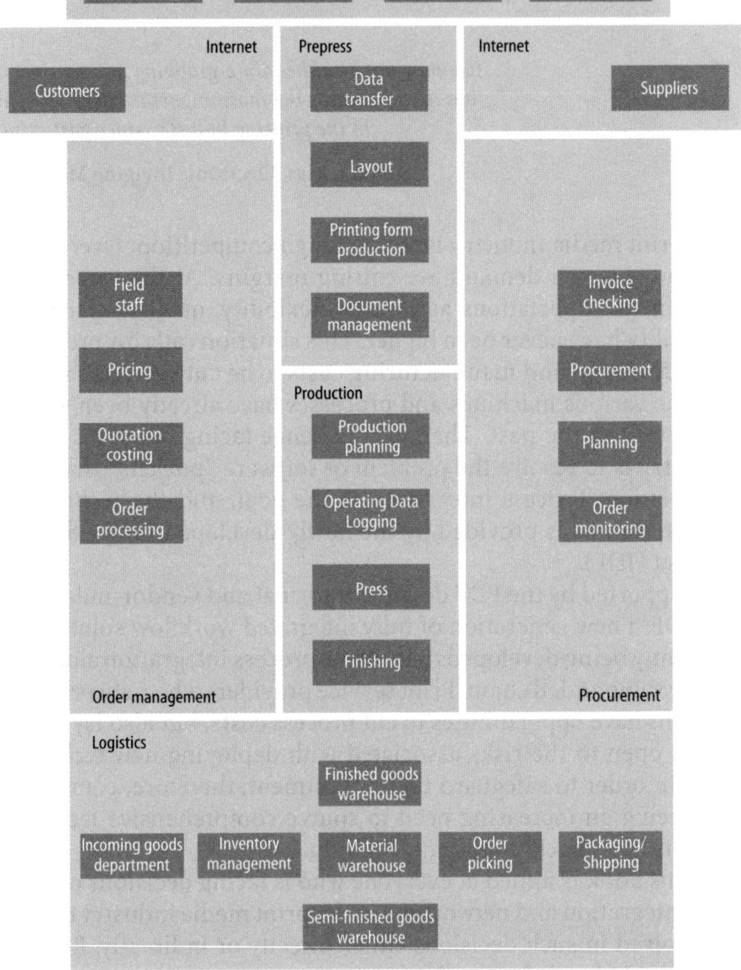
Internet
Customers
Prepress
Data transfer
Layout
Printing form production
Document management
Internet
Suppliers
Field staff
Pricing
Quotation costing
Order processing
Order management
Production
Production planning
Operating Data Logging
Press
Finishing
Invoice checking
Procurement
Planning
Order monitoring
Procurement
Logistics
Finished goods warehouse
Incoming goods department
Inventory management
Material warehouse
Order picking
Packaging/ Shipping
Semi-finished goods warehouse

2 Process integration in the print media industry

The print media industry is going through a period of upheaval. Innovations and industrialization concepts that have already been implemented in other industries have recently started to find their way into this sector too, while IT terms such as Enterprise Resource Planning (ERP), Management Information System (MIS), Computer Integrated Manufacturing (CIM), Customer Relationship Management (CRM), Supply Chain Management (SCM) etc. are also beginning to make their presence felt.

The arrival of the Job Definition Format (JDF) in the year 2000 marked a major step forward. This new standard paves the way for the development of fully integrated process systems that promise a high degree of transparency, flexibility, quality and efficiency. The networking of print service providers via common standards will provide the next major development push for the industry following the establishment of an end-to-end digital workflow in prepress.

JDF

The Job Definition Format (JDF) describes and collects all the relevant data and process steps for a printing job. JDF is based on XML and is maintained and developed by the CIP4 consortium. www.cip4.org.

Multiple media changes occur between the different departments of a print service provider such as order management, accounting, production and logistics, and also vis-à-vis customers and suppliers. Even activities within individual departments are often performed using different software applications that cannot communicate with each other. These inefficient processes make their mark on process costs. The goal of process integration is to cut these costs. This is done by transferring data generated in the individual departments seamlessly across various networking routes.

2.1 The essential problem with non-networked production – process costs

Non-networked print job production generally consists of a series of isolated software solutions and changes of media that lead to lack of transparency and inputs being duplicated. This in turn results in unnecessarily high processing and communication input when managing orders. The more complex the jobs and the shorter the runs, the greater the impact these inefficiencies have, resulting in higher process costs.

Process costs are costs incurred in the process of manufacturing a print product and include indirect costs deriving from order processing, job preparation, production, logistics, etc. In simplified terms, the costs incurred in order management, job preparation, prepress and logistics are proportionate to the number of jobs and their complexity, while in press and postpress, factors relating to the size of the print run have most influence. Investing in suitable platesetter (Computer-to-Plate) technology and state-of-the-art printing presses can reduce process costs in production. Process costs in order management, job preparation, logistics and prepress can be reduced by using "lean" workflows which ensure inputs only have to be made once and that different departments do not have to get involved in complex coordination procedures.

A study conducted in 2000 by the Rochester Institute of Technology found that inefficient processes become a problem when average print run sizes fall. According to this study, in the years 1998 to 2000 alone, the number of print jobs with runs smaller than 2,000 increased from 28% to 44%.[1] This leads to increased order costs, above all in prepress, proofing and order management, i.e. sectors that are not governed by print run levels.

This is confirmed by research by Heidelberger Druckmaschinen AG. A study from 2000 found that, on average, 24% of job costs are incurred in the order management sector. Prepress accounts for a further 23% of costs. Print service providers anticipate that costs in these two sectors will rise to 28% and 26% respectively in future. Indirect administrative costs and prepress costs that are not governed by print run levels are further eating into margins. Production will become unprofitable if countermeasures are not taken.

[1] Romano, Rochester Institute for Technology, 2000

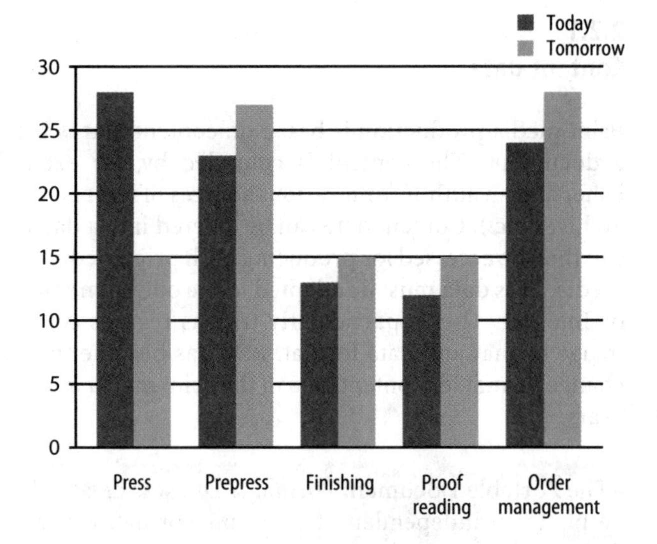

Figure 1
Manufacturing costs as % of order costs. Source: Heidelberger Druckmaschinen AG

2.2 Data types in the print media industry

Data types store technical and commercial information in digital form. The structure of the different data types varies depending on what they are used for. Some have already been standardized, others have not. In addition, interpersonal communication is still required at those interfaces where communication is not supported by technology.

The data encountered in the print media industry is discussed below. It can be subdivided into:

- Content data,
- Master data,
- Job data,
- Production data,
- Control data,
- Operating and machine data,
- Quality data.

It can be difficult to separate these data types in individual cases, since the data can be used in different contexts.

2.2.1
Content data

Print media production is based on content that is processed into a document. The content is compiled by, for example, a publisher and an author from various sources of information (agencies, archives etc.). Content data can be entered into a database where it can then be accessed for producing both print media and electronic media. This data must be adapted to the output medium, e.g. paper or Internet. This is particularly true as regards color, resolution, image format and data format. PDF has become the standard exchange format for content data in the print media industry in recent years.

PDF

The Portable Document Format (PDF) was created by Adobe as a platform-independent data output format. It is a recognized standard in the print media sector, and has also been widely adopted in the IT sector.

2.2.2
Master data

Master data is data that is needed again and again in all areas of order management, planning and production. It includes product structures, equipment, human resources, production structures, and customer and supplier addresses. Master data does not change a great deal and can be used over and over if it is stored in a database. It is best to maintain master data at a central location and make it available to staff at their workstations through access rights.

2.2.3
Job data

Job data describes a job. Traditionally, job data is transferred to the individual workstations in the form of a job ticket. The job ticket serves to ensure that the job proceeds smoothly through production. The job name and number, customer name and number, etc. are fixed and do not change.

Job data is created by entering it into the order management system no later than when the order is awarded, but usually as early as the quotation phase.

JOB TICKET — No.: 04-0013

Delivery date:

Customer:			
Heidelberger Druckmaschinen AG	Cust. No. 1	Reference:	prinance_dba
Kurfürstenanlage 52-60	Phone. 1:	Last order no.:	
	Fax:	Date:	28.03.2004
69115 Heidelberg	Mod.:		

Notes: **House Colors: Blau HKS 42 = Cyan 100, Magenta etc. Always 5 print samples in the job ticket!!!**

20,000 brochures, summer festival 2004

Extent:	4 pages	Form number:	
Finished size:	21 cm x 29.7 cm	Order number:	
Open size:	42 cm x 29.7 cm		

[x] New print   [ ] Reprint   [ ] Reprint w. changes

Sheet(s):	Front colors:	Back colors:
1 sheet of 4 ups 4/4-colored	Euroscale	Euroscale

Composition/Repro:			Correction: [ ] YES [ ] NO	
Correction until:			1st correction:	
Correction to:			2nd correction:	
Composition ready until:			3rd correction:	
			O.k. to print at:	
2	DTP	Composition work		
2	COLOR SCAN	4 Scans 4c A4	60 lcm screen	
Form Production:			**Archive of assembly:**	
1	CTP	Output computer to plate	/ Plate 103 cm x 77 cm	4
Print:			**Archive plates: [ ] YES [ ] NO**	
1	CD102-5L	Set up		4
1	CD102-5L	Printing: Work and turn		11,000

Figure 2
Typical job ticket

Order management system

Order management systems are software applications that enable orders to be costed and processed. In the print media industry, such systems are known as costing software or industry-specific software.

Customers are now increasingly able to submit job data to the order management system online via Internet portals.

2.2.4 Production data

Production data defines the production process between different software applications and machines. It is mainly generated in the job preparation and prepress stages and passed on to downstream production resources. Production data is not associated with specific machines and is therefore still relatively flexible. This data is encoded in industry standards such as ICC profiles and the Print Production Format (PPF), in quasi-standards such as the Portable Job Ticket Format (PJTF) or in proprietary formats.

PPF, PJTF

Numerous parameters for a print product can be described in prepress in Print Production Format (PPF) and used later to preset presses and finishing machines. PPF was stewarded by the CIP3 Consortium, which was also responsible for its ongoing development. http://www.cip4.org

Portable Job Ticket Format (PJTF) is a format from Adobe for automating the prepress workflow. The PJTF set of functions is included in JDF. www.adobe.com

Thanks to the intensive cooperation between agencies and prepress, production data has already been widely standardized since the introduction of Postscript in the mid-1980s. The possibilities opened up by remote proofing and PDF certification will also close the last remaining gaps in the automatic workflow between agencies and print service providers.

2.2.5 Control data

Production data can be transformed into control data using machine and device control software. Calibration curves, ICC profiles and folding type catalogs are among the means available for doing so. Production staff can use production data directly or in slightly

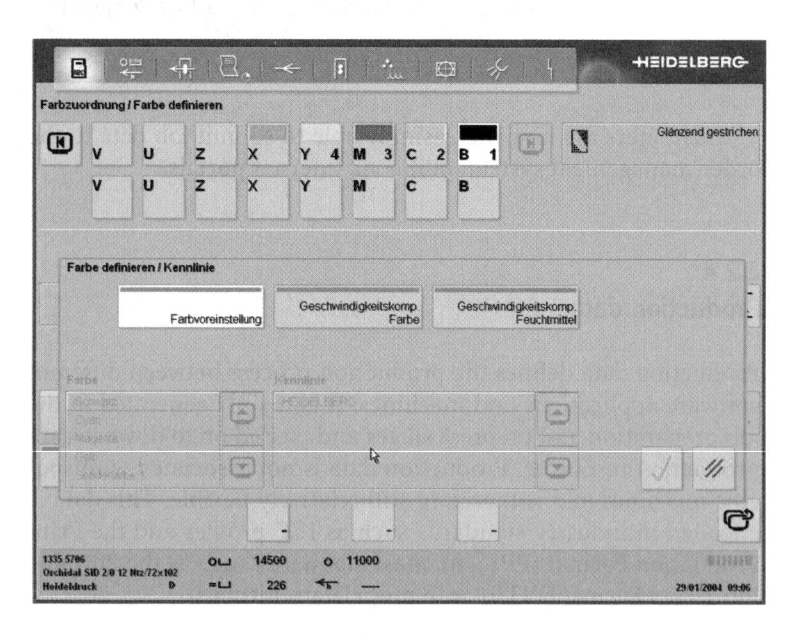

Figure 3
Input of control data at the CP2000 Center

adapted form to steer processes at the control consoles. In addition, presses, for example, require various items of data that do not come from prepress, but must be input directly at the control console. These include printing unit assignment, application of dampening solution, powder length, etc.

2.2.6
Operating and machine data

Operating and machine data is needed to evaluate the processes actually carried out for production control purposes and actual costing. Evaluating this data provides information about the status, utilization and availability of production resources. Machine data is technical data that is loaded directly from the machine or workflow and must be interpreted in business management terms.

Operating data is traditionally recorded by staff on daily dockets and provides information about resources used and materials consumed. In addition to job-related data, non-job-related data such as servicing times, training times, etc., is also recorded.

2.2.7
Quality data

Quality data is data that serves to maintain a defined quality standard or ensure continuous production. It includes densitometric and photospectral measured values and information on the chemicals necessary for plate development. Quality data is important to print service providers as it helps them ensure high production reliability within predefined tolerances.

Quality data can be used directly to regulate the production process. If a measured value exceeds the required tolerance zone, the press has to be adjusted or a warning is issued. Quality data is also required for documentation. Print service providers aiming for ISO certification in particular must document – and be able to provide retroactive proof of – the levels of quality they have attained.

2.3
Networking routes of process integration

Data is exchanged within an application or between different applications at print service providers in digital, written or oral form.

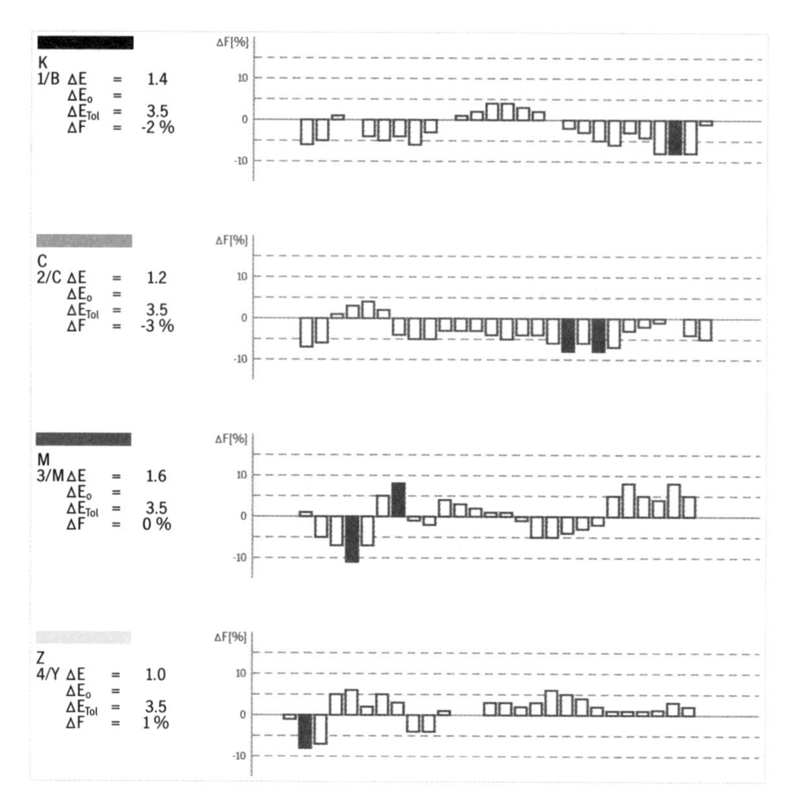

Figure 4
Report from a color measurement and control system

A stream of data interconnects various applications to form networking routes.

Due to the high division of labor in the print media industry, data streams are not limited to one site or one company. For example, customers make inquiries, approve the print result, and ask when production will be ready for delivery. Suppliers offer their products with price lists and product information. Agencies exchange content data with print service providers.

Integrating the processes used by print service providers is a major task. Networking requires an investment in up-to-date, network-enabled software applications and services. To be able to communicate in the network, older machine controls have to be brought up to the current level of hardware and software. Last but not least is the not inconsiderable work required internally to introduce networking and train staff.

However, print service providers do not have to be networked in every aspect, nor does networking have to be completed in one fell swoop. It is therefore a good idea to split up the complete package of networking possibilities into individual networking routes, then focus on particular areas of investment. Specifically, we can identify the following networking routes:

- E-business,
- Job preparation,
- Machine presetting,
- Job planning and control,
- Operating data loggingand actual costing,
- Color workflow.

Depending on the order structure and the machine park, analyses are needed to identify which networking routes actually offer feasible benefits for the particular print service provider.

2.3.1 E-business

The purpose of the e-business networking route is to minimize process costs between customer and print service provider, increase customer loyalty through solutions specifically tailored to customers, and offer new services. Since they enable 24/7 availability without regional limitations, e-business solutions offer print service providers the opportunity to expand their market.

24/7 availability

24/7 availability (24 hours, 7 days) refers to the availability of the service to remove faults and provide help with user problems.

E-business solutions have begun to penetrate the market, especially in the prepress and digital printing sectors. In prepress in particular, this has led to a widening in the range of services offered, with providers supplying services such as image databases, standard print products and coordination of content data. By its very nature, digital printing is ideal for very short runs, and must therefore grapple with relatively high process costs in order processing, which can be reduced using Internet portals.

Key aspects of the e-business networking route include:

- E-business solutions must be marketed. Potential customers have to be actively approached. Given the relatively large training and advertising effort required by e-business solutions, target customers should generate comparatively high sales to justify the effort on both sides.

- The degree of integration with the order management system is key for ensuring low process costs. The fewer interventions required to process orders, the higher the probability that the e-business model will be profitable.
- E-business solutions should be particularly adaptable to customers' needs. An unwieldy user interface, violation of security guidelines, or the need to involve companies' IT departments can quickly jeopardize e-business projects.
- The Internet portal provider becomes a 1st level service for its customers. The round-the-clock availability of the portal must be ensured. Companies should consider whether to operate the Web server themselves or look for a provider whose services they can use.

1st level service

1st level service is a service offered to customers giving them an initial point of contact for user questions and problems. For more complex questions, 1st level service can seek assistance from 2nd level service.

The task of defining the job in hand is transferred to the customer's sphere of responsibility. This significantly reduces the administrative effort required for order processing, and clearly increases customer loyalty, particularly where the solution is integrated in the customer's ERP system.[2]

2.3.2 Job preparation

Order management involves creating job tickets, which are then augmented with production data, print samples etc., and transferred to prepress, the printroom and finishing operations. Job-related information from all the cost centers in the production process is described in the job ticket. Changes during the production process are input in the job ticket and reported back to order management. Networking the job preparation stage is intended to eliminate the change of media between order management and production. Further aims include the avoidance of duplicate inputs, more up-to-date job tickets and increased reliability in production.

[2] An overview of e-business can be found in König: E-Business@Print, Springer-Verlag, 2004.

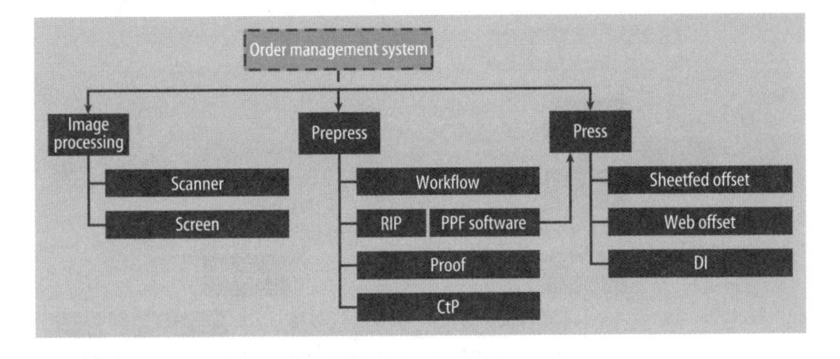

Figure 5
Job preparation networking route

Key aspects of the job preparation networking route include:

- Standardized job designations at the print service provider or within a production network simplify communication and help to avoid errors, especially in cases where job preparation data is sourced from different places.
- Complete, transparent and up-to-date job tickets are important for ensuring error-free production at every workstation round the clock. Information should be reduced to what is actually needed and must be easy to identify clearly, for example with the help of previews.
- Job preparation stations have become established in prepress, press and finishing stages. They increase the degree to which production resources are utilized and enable consistently high quality even when staff have different levels of training.

Thanks to networking, operators are immediately notified of any changes in the job or processing caused by e.g. a press becoming unavailable at short notice. The modified electronic job ticket is immediately adapted to this end at the corresponding control consoles.

2.3.3 Machine presetting

Machine presetting aims to reduce setup times and waste in production. As the name suggests, presettings must normally be carried out on the machines. The area coverages determined in prepress are converted into ink zone values that are used to control the individual ink zone valves on the press. In addition, the autoregister, cutting, folding and saddlestitcher marks created in the impositioning software can be employed for presetting presses and finishing machines using the PPF file described above.

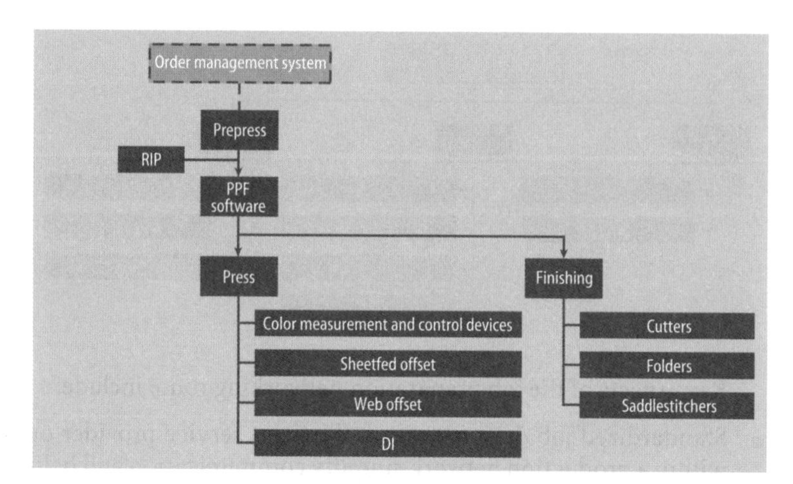

Figure 6
Machine presetting networking route

Key aspects of the machine presetting networking route include:

- The necessary knowledge about the production process must be available in order management and prepress. To maximize reliability in production, the parameters that can be selected in the order management system and in the print workflow should be restricted to those that are actually available. Incorrect presettings, far from saving working time, just lead to problems.
- In order to avoid presettings being wrongly allocated in production, it is important to make sure what job is being processed. Either the job number must be entered manually in prepress, or a rudimentary networking with order management must be established.
- In order for the investments in machine presetting to pay off, the presetting options for the machine in question must be comprehensive enough to cut working time significantly.

CIP3

The CIP3 (International Cooperation for the Integration of Prepress, Press, and Postpress) Consortium was responsible for maintaining and disseminating the Print Production Format (PPF). www.cip4.org

Machine presetting was made possible in 1997 thanks to the manufacturer-independent PPF format of the CIP3 Consortium, and became established at the latest in 1999 with the increasing popularity of CtP. Other machine presettings are based on proprietary formats, such as paper and sheet information from the order management system and is used to preset suction air and printing pressure at the feeder and delivery.

2.3.4 Production planning and control

Production planning and control aims to ensure optimum use of production resources while keeping to specified delivery dates. This requires that activities captured in order management are structured, and carried over into production plans for prepress, press and finishing. Defining the production sequence accurately is important for keeping to delivery dates, especially where it is not possible to process jobs sequentially. An important task for planners is to optimize the utilization of the press. The printing units must be allocated so as to minimize downtimes due to ink changes. In addition, the printer must be relieved of unscheduled activities such as looking for plates, paper and ink.

Key aspects for the production planning and control networking route include:

- Planning changes must be quickly forwarded to all affected workstations. Many orders come with only short notice, making it necessary to change the production plan many times each day. The various operations must be dovetailed such that, when changes occur, the schedule is updated automatically and conflicts are avoided.
- There should be a system in place for planning in maintenance operations and periodicals. Not all jobs are pre-costed, and it must be possible to add such jobs into the production planning system and to factor in times that are not job-related.

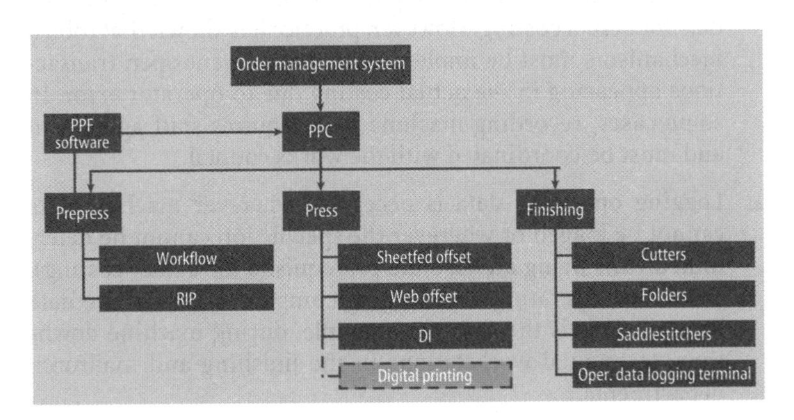

Figure 7
Production planning and control networking route

- An electronic planning system must map the entire production process from prepress to delivery. It should also be possible with database support to perform both rough and detailed planning separately, and to plan printing and finishing separately. The planner must have information on the availability of consumables such as plates, paper and ink. Ideally, it should also be possible to implement a networked planning system across multiple sites so that partner companies and suppliers can also be integrated.

While manual planning and simple spreadsheets are widespread, less use is made of the capacity planning features of the order management system. The networked production planning and control system that enables data to be passed on in real time is a true rarity.

2.3.5 Operating data logging and actual costing

Operating data logging and actual costing deliver the figures used for producing the statistics required for in-house controlling. While working times and material consumption levels are paramount in prepress, interest in the printroom and finishing operations is mainly focussed on which cost unit has been utilized and for how long. The print service provider is guided by these key figures. Basically, the more data is logged, the better the processes can be monitored.

Key aspects of the networking route for operating data logging and actual costing include:

- Exporting machine data is a quick way to log reliable up-to-date data for actual costing. However, practice has shown that robust mechanisms must be implemented that prevent open transactions appearing in the actual costing due to operator error. In some cases, recording machine data requires staff agreement and must be coordinated with the works council.
- Logging operating data is necessary wherever machine data cannot be logged or wherever the specific job cannot be determined (this being an absolute prerequisite for actual costing). Therefore, operating data logging complements machine data logging. This is the case, for example, during machine downtime, at manual workstations in the finishing and mailroom departments.

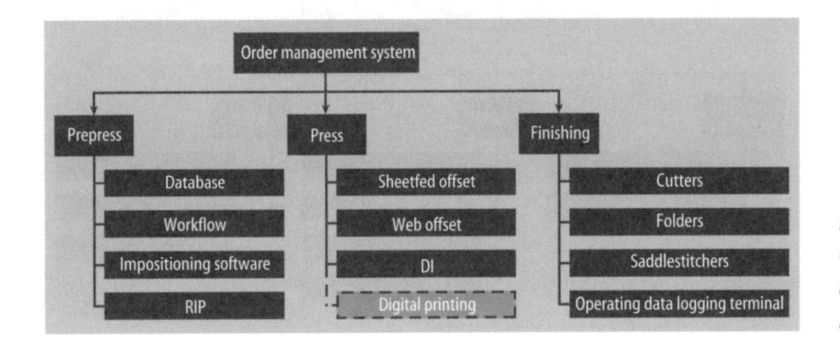

Figure 8
Operating data logging and actual costing networking route

- It also makes sense to network the order management system with the financial accounting system so that payment transactions with banks, payroll operations and annual accounts can all be processed automatically. Payroll is usually part of the financial accounting software, which receives the necessary data from the order management system via interfaces.

In recent years, special operating data logging terminals have become increasingly popular. In the printroom, machine and operating data logging has now matured sufficiently to allow informative actual costing with inputs at the control consoles. In prepress exporting machine data has only been made possible by the most recent generation of workflow systems.

2.3.6 Color workflow

It's not only global companies with multiple sites who face the challenge of ensuring color stability from the digital copy right through to the final print. Securing predictable, constant color quality is also absolutely essential for small companies if they are to remain competitive.

This challenge requires color management, which produces a standardized visual impression on the various output devices (scanner, screen, proofer, press). It makes (virtually) no difference what materials, devices and machines are used, as long as they have been calibrated and profiled.

Key aspects of the color workflow networking route include:

- The print service provider must be fully prepared to standardize his workflows, as this is the only way to ensure stable processes with a reasonable outlay. A few standard workflows should

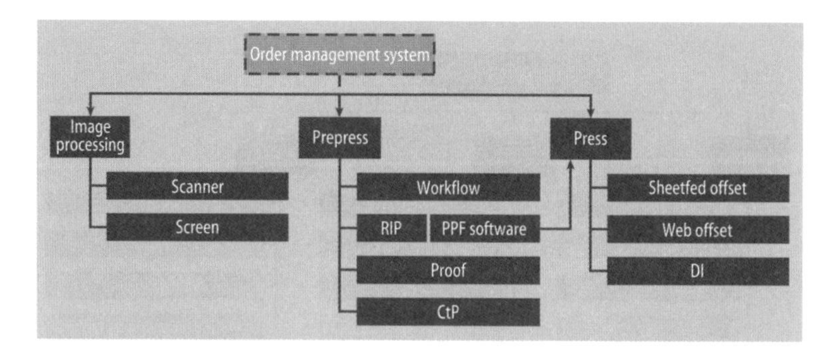

Figure 9
Color workflow networking route

be chosen rather than multiple production resources and processes.

- In order to implement a networked color workflow, it is also essential that these workflows are consistently adhered to. Consistent color quality can only be achieved if the conditions are kept within the tolerance range. Company guidelines regarding calibration, ink, paper, and exchanging blankets are to be strictly adhered to. Any process deviations that occur must be rectified at their source and must not be compensated for by the printer.
- The success of the networked color workflow can be measured in terms of a reduction in waste paper and complaints, rather than "better" printing. The likelihood of success is strongly dependent on the circumstances at the outset and the company's range of products.

Connecting spectrophotometric measuring devices, which measure the entire sheet or the color control bar, with the reference values from prepress opens up interesting possibilities. The digital reference values are compared with the measured values and the press's ink zone valves are adjusted accordingly.

2.4
Why has process integration failed so often in the past?

The networking routes presented above exhibit significant potential for optimizing the production process and cutting process costs. Attempts have long been made to network print service providers. The providers Creo with Networked Graphic Production, Heidelberger Druckmaschinen AG with Prinect, MAN Roland with PeCom and

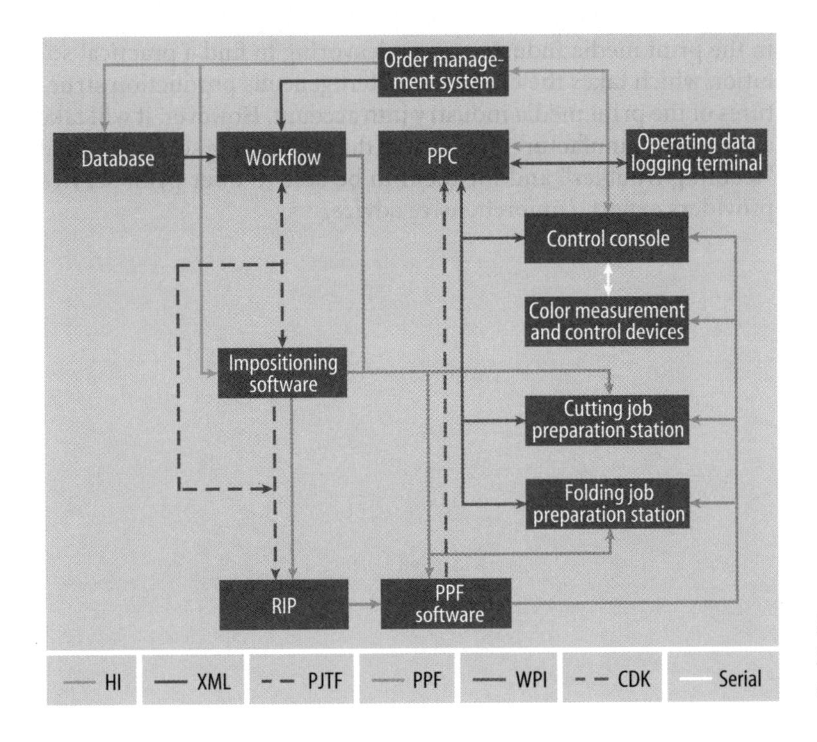

Figure 10
Multiple interfaces of a networking solution

the PrintCity consortium have for some time been offering solutions that enable order management to be networked with production.

However, apart from ink zone presetting of presses, these attempts have not yet met with any great success. This is mainly due to

- the insufficient market-readiness of various integration scenarios,
- too many proprietary, incompatible interfaces,
- the significant integration, testing and installation outlay required by the numerous proprietary interfaces,
- a poor cost/benefit ratio when networking different makes of product,
- the interfaces' limited functionality that does not enable any meaningful evaluation and presetting,
- the lack of a central networking architecture with consistency checking.

On the technical level, these weak points are set to be eliminated with the establishment of the Job Definition Format. Manufacturers in the print media industry are endeavoring to find a practical solution which takes the extremely heterogeneous production structures of the print media industry into account. However, it will take a while for manufacturers to correct the networking solutions' final "teething troubles" and for them to be able to offer print service providers expert, comprehensive advice.

3 Job Definition Format

The concept for the new JDF industry standard was presented at the Seybold Conference in early 2000 by Adobe, Agfa, Heidelberger Druckmaschinen AG and MAN Roland. The definition of JDF has its origins in the idea of creating a standardized, manufacturer-independent and comprehensive data format for the print media industry.

- JDF enables business management aspects to be seamlessly dovetailed with production-oriented aspects,
- JDF provides a clear, fully-integrated structure for the entire workflow,
- JDF enables end-to-end production control, even across corporate boundaries,
- JDF creates transparent workflows by logging the relevant target and actual data,
- JDF is based on the existing formats PPF, PJTF and IFRAtrack,
- JDF is Internet-compatible.

JDF enables both the horizontal integration provided up to now by PJTF and PPF and vertical integration between order manage-

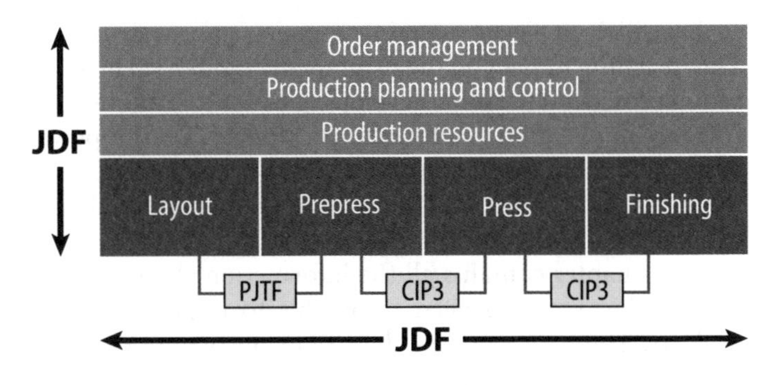

Figure 11
The JDF workflow

ment, production planning and control, and production resources. The many existing standards are combined in one single industry standard.

JDF also has an integrating effect on the different market segments of the print media industry. For example, IFRAtrack, used for messages in newspaper printing, is also included in the new standard. Whereas in commercial printing manufacturer-specific standards were prevalent in the past, in future messages in both newspaper printing and commercial printing will be transmitted via a common standard.

IFRA

The origins of the name IFRA go back to the "INCA-FIEJ Research Association", where "INCA" stands for "International Newspaper Colour Association" and "FIEJ" for "Fédération Internationale des Editeurs de Journaux". www.ifra.com.

The establishment of a uniform, comprehensive standard has significantly reduced the effort required to develop and implement networking solutions. Print service providers are able to deploy software applications from different manufacturers without having to make any compromises when it comes to networking.

JDF is maintained and developed by the CIP4 Consortium. The CIP4 Consortium currently consists of more than 200 manufacturers, organizations and print service providers who are driving the standard forward and have committed themselves to bringing JDF-compatible products onto the market.

CIP4

The International Cooperation for the Integration of Processes in Prepress, Press, and Postpress (CIP4) organization has taken on the task of maintaining, developing and disseminating JDF. www.cip4.org

Like PDF, JDF is provoking a major upheaval that will have a far-reaching impact on the print media industry. The JDF specification provides a basis for developing new, networking-compatible software applications and machines. Today, JDF is a powerful yet unfinished standard. Numerous working groups are still busily fine-tuning various aspects of JDF with the goal of optimizing its range of applications.

This chapter contains all the information about Job Definition Format that you will need to be able to implement networking projects. The issues of how the CIP4 Consortium can implement a uniform standard and its interplay with other standards influencing the print media industry are also interesting. Finally, the frequently asked question regarding the security of the Job Definition Format will also be clarified.

3.1 Who needs to understand what about JDF?

JDF is a complicated matter. The description of the standard in specification 1.2 already takes up over 700 pages, which are sure to increase as the standard continues to be developed. This gives rise to the question of who needs to understand what about JDF.

- *Software developers* require an in-depth knowledge of the JDF structures in order to design state-of-the-art software structures and products that measure up to JDF workflows.
- *Programmers* must be able to either use the specification to program a JDF interface or employ the libraries supplied by the CIP4 Consortium.
- *Users* should be able to use JDF-based solutions without having to understand the details of JDF.
- *Decision-makers* need background knowledge about JDF to support their investment decisions. This knowledge should enable them to critically examine the networking solutions on offer.
- *Consultants* must be able to understand the interplay of JDF interfaces from different suppliers and suggest a suitable networking architecture tailored to the specific operation. Consultants must also test the suitability of individual software applications within the networking concept and be able to design a viable networking concept.

3.2 JDF basics

Much has been written on the topic of JDF recently, yet ambiguities remain. Is JDF a standard, an architecture or a software application that enables print service providers to become networked? Is JDF a standard and therefore manufacture-independent or are there manufacturer-specific versions of JDF? The past four years have seen expectations of what JDF can do grow to unrealistic proportions in some quarters, and these must be corrected. To this end, here is a brief description of what JDF can and can't do:

- JDF is a standardized data exchange format enabling jobs to be specified and handled in detail, and as such the Job Definition Format has a job-related setup. Data that is not job-related can only be depicted indirectly, if at all.
- JDF is based on XML and as such applies no restrictions to the field lengths of communicated data. This is not a limitation of JDF, but does cause problems for some software providers whose older products do have such restrictions.
- Since JDF is relatively new, there are understandably still a few restrictions and incomplete definitions that will be gradually eliminated in future versions.
- JDF is not a software architecture, nor does it define any such architecture. Neither is JDF a software product or database, and it is most certainly not a user interface with predefined entry fields.

XML

eXtensible Markup Language (XML) is a meta-language standardized by the W3C and used for defining document types. This extensible meta-language is platform-, application-, language- and domain-independent. The W3C was founded to develop protocols for the World Wide Web that ensure the problem-free exchange of data. www.w3c.org

Not many suppliers can afford to develop completely new JDF workflows in one fell swoop. The next few years will therefore see many products that still make only partial use of the possibilities of JDF. This makes investment decisions more difficult, since buyers will most likely find it hard to gain a clear picture of the degree to which JDF is actually supported.

3.2.1 Interfaces

Significant standardization on several IT levels is required for JDF data exchange to function reliably. The OSI 7-layer model illustrates this system.

"Open Systems Interconnection" (OSI) is a working group of the "International Organization for Standardization" (ISO) charged with preparing standards and protocols for data exchange in digital networks.

OSI

Each of the layers is to be implemented independently so that changing the mechanism on one layer, for example replacing copper cables with fiber optic cables, will have no impact on the layers above and below it. This yields a compatible, future-oriented interface architecture.

Standard interfaces based on hardware standards and standard protocols enable reliable communication and make it easier to connect components. Recent years have seen the establishment in the office sector of networks based on Ethernet and TCP/IP, and users today expect to be able to simply plug their PC in at the socket, set a few parameters, and then have a fully functional network connection.

The Transmission Control Protocol/Internet Protocol (TCP/IP) is used to transmit data in distributed networks. IP makes sure that data packets are forwarded from node to node with the help of IP addresses. TCP ensures that data transfer is correct and complete.

TCP/IP

The system described also applies to software interfaces. Today cutting-edge technologies such as XML provide good possibilities for defining universal interfaces based on a standardized foun-

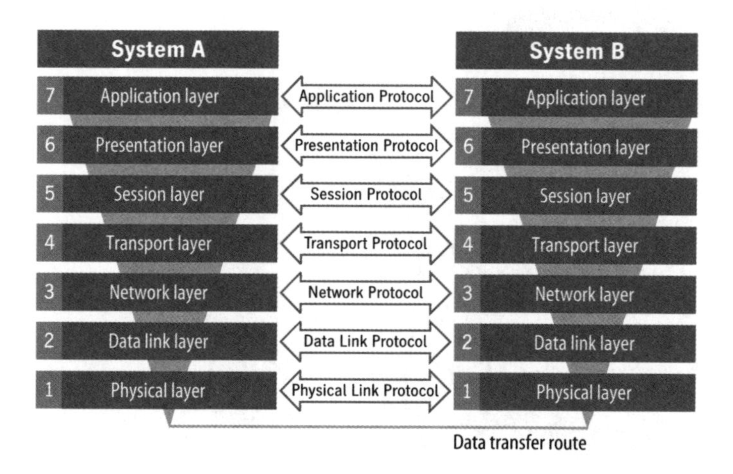

Figure 12
Layer model of the communication level

dation. The XML files can then be sent via standardized transfer media.

3.2.2 How does JDF work?

Job Definition Format is characterized by a tree structure in which the product and process structure of a job can be mapped in the hierarchical relations between individual JDF nodes. The starting point is the customer's product description as contained in the inquiry or order. For example, a DIN A4 brochure with 16 color and 4 b/w pages is to be produced in a run of 2,000 copies. The digital content data is to be supplied.

The tree structure also enables serial, parallel, overlapping and iterative processes to be mapped in any desired combination across distributed sites. The JDF elements required for this and the interplay between them are presented below.

Master data, such as the address, contact, telephone number and e-mail address of customers is usually stored and maintained together with a detailed customer profile in a database. In JDF, the relevant customer data for the job in question can be included in a JDF element called *CustomerInfo*. This data is then available when drawing up a quotation, coordinating a check sheet or delivering the finished product.

The *product structure*, which mirrors the customer requirements, describes what is to be produced. It includes specifications describing what grade of paper, print quality, type of binding, size of run, format etc. are required. A rudimentary product description can be created using a customer inquiry from e.g. an Internet por-

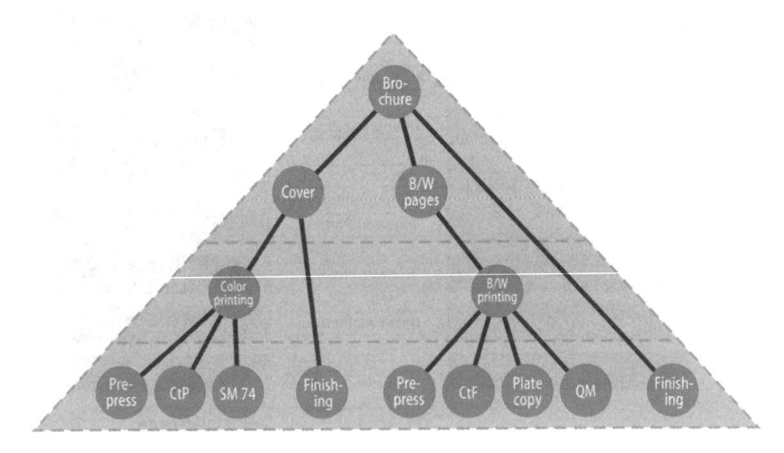

Figure 13
JDF tree

tal and transferred to the order management system as JDF. This means that the data is already available in structured JDF form in the event that the order is placed. Using the product description, a tree structure is built up in JDF. The product and its constituent products are designated in the product structure as product nodes. For standard products that are produced frequently, JDF templates with corresponding product node structures can be archived and made available to the user.

In order management, the product description is developed into the *process structure*. The process structure describes how the product is to be made. This is done in order management by establishing which scans and proofs are needed, which plates are produced, and which machines are to be used in which order for printing, cutting and folding. Depending on how the work is allocated, the job preparation department and the planner further refine the process structure. All the processes necessary to make the constituent products are defined and arranged in a production sequence. The production segments are also represented as nodes in the process structure. Individual processes such as folding, binding and cutting can be assigned hierarchically to specific process groups, for example a finishing group. To reduce the amount of work, the process description should best be loaded as an existing JDF template and modified where necessary. If required, it can of course be created from scratch.

Various *process resources* such as ink, paper, coating etc. are deployed for the individual production processes. These produce new process resources such as OK sheets and waste, which are in turn required as input for further processes. In the JDF tree structure, these are known as *resources*, which in JDF also includes digital material such as PDF files, machine parameters and ICC profiles that are referenced in the JDF. Some processes can require approval before continuing. After a proofing process, for example, a (digital) signature may be required from the customer as a *resource* for the following process.

The *process workflow* needed to manufacture the product is laid down by the JDF tree structure. Pointers describe the path that resources take through the tree structure and in doing so define the production sequence. In JDF, these pointers are described as *ResourceLink* elements.

The *process description* is deposited by order management and job preparation in a *NodeInfo* element. The *NodeInfo* element is included in every node in the JDF tree structure and contains all the logistical information required for the execution of the node. This primarily relates to precise process times. Among other infor-

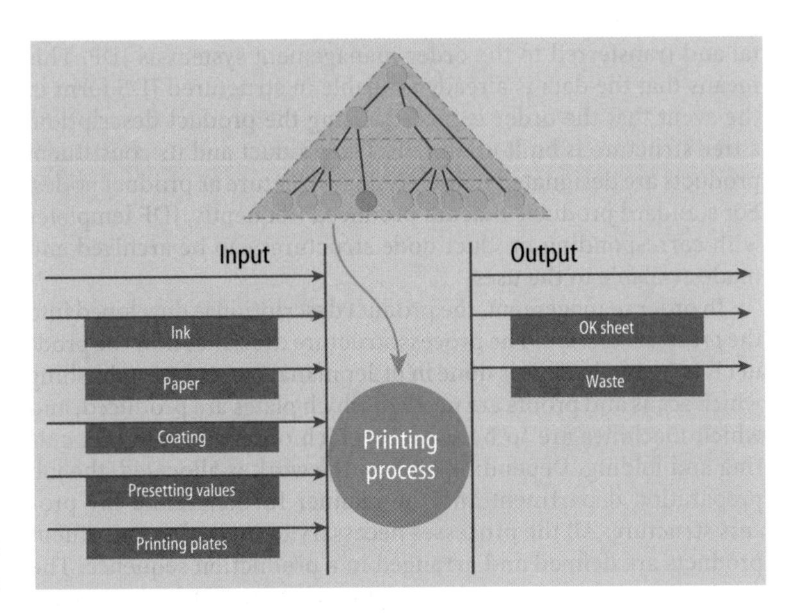

Figure 14
Process resources

mation that can be specified in the *NodeInfo* are the consequences that e.g. overrunning the time will have for the rest of the process workflow. It also defines which machine or member of staff will carry out the process. Thanks to JDF's modular design, the process description can be modified at any time.

After completion of every processing stage, the relevant key figures are logged in JDF. The *AuditPool* element, which is appended onto the relevant JDF node, is responsible for recording this data within the JDF tree structure.

The following are among the events that can be recorded by *AuditPool* elements:

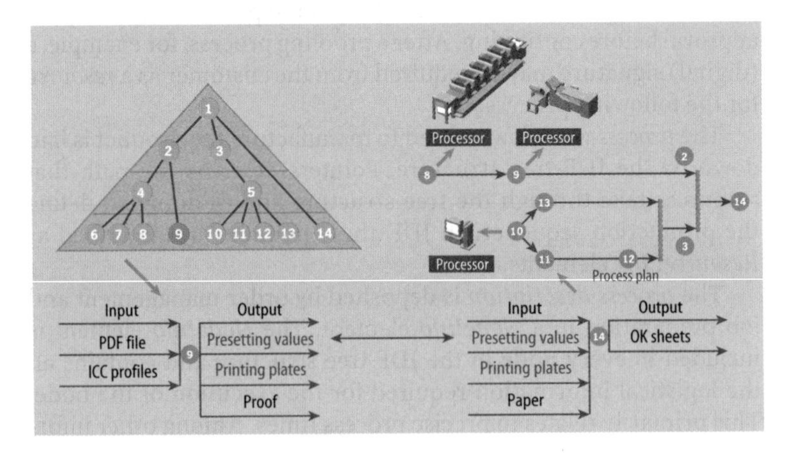

Figure 15
Process workflow

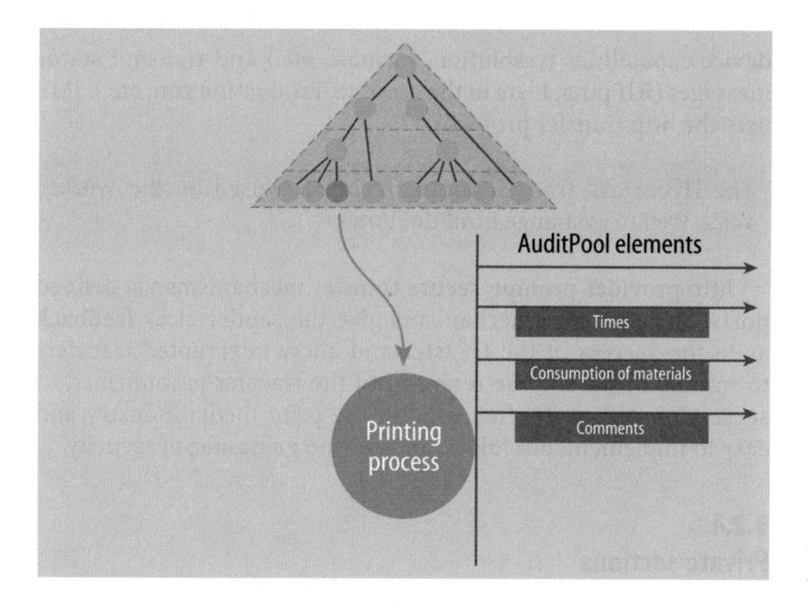

Figure 16
AuditPool elements

- Creation of, and changes to, a JDF node,
- Current data on the production workflow and consumption of resources,
- Availability of resources, spawning and merging of *Resources*,
- Errors in the JDF identified during editing and simulation, such as unnecessary *ResourceLinks*, incorrectly linked *resources*, missing *resources* or missing links, that were identified by a messaging element or during a test run,
- Process start and end times, setup times, interruptions and staff changeovers,
- Unscheduled events and problems executing a process node.

AuditPool elements are used for documentation and cannot be changed after the process has been abandoned or completed. They can however be read and evaluated at any time.

3.2.3 Job Messaging Format (JMF)

The Job Messaging Format (JMF) is used to transfer messages and complements the Job Definition Format. Up-to-the-minute information defined using JMF can be exchanged while production processes are in progress. Job lists are created using JMF to report on device capabilities (resolution, formats, etc.) and transmit status messages (RIPping, Plate in the cassette, Production run, etc.). JMF uses the http transfer protocol.

http

The Hypertext Transfer Protocol (http) is used on the World Wide Web to exchange html documents.

http provides prompt, secure transfer mechanisms via defined ports. Such transfer mechanisms give the sender clear feedback as to the success of the transfer and allow interrupted transfers to resume or parts to be resent until the transfer is confirmed as successful. Although often used in the print media industry and easy to implement, hot folders provide no guarantee of security.

3.2.4 Private sections

Private sections enable additional manufacturer-specific functions not specified in the JDF standard to be encoded and transferred within a JDF.

Private section

Private sections are manufacturer-specific, XML-based encodings that are not included in the JDF standard. Other manufacturers are generally unable to read or process the information stored in the private sections.

Private sections can be used to temporarily fill in any definition still missing from the Job Definition Format. This enables functions that have yet to be included in the standard to be implemented within the JDF tree structure until they are included in the JDF standard.

Some software applications also misuse JDF as packaging for proprietary interfaces that are then encoded in a private section. Potential buyers should be wary of such products, given that they offer no guarantee of compatibility with other products that adhere to the standard. A JDF workflow must therefore be able to ignore the private section part without this causing the process to be aborted.

3.3 Implementation of JDF by CIP4

A great deal of standardization is required to get JDF accepted in the print media industry. The CIP4 Consortium has therefore adopted a number of measures to, on the one hand, support manufacturers' efforts to develop JDF interfaces and, on the other, prevent subsets of JDF that undermine JDF's effectiveness as a cross-manufacturer standard.

The CIP4 Consortium makes free open source libraries for the programming languages C++ and Java available to its members. These libraries are intended to make it easier for manufacturers' software developers to write JDF applications, reducing development times and therefore cutting costs. The use of common libraries is also a means of ensuring that JDF is always interpreted consistently.

In a move designed to identify deviations from the JDF standard and counter any tendencies toward non-standard forms of JDF, the CIP4 Consortium and the Graphic Arts Technical Foundation (GATF) signed an agreement on the certification of JDF solutions in January 2004. In line with this agreement, procedures and tools for the certification of JDF interfaces are being developed hand in hand with the 'Interoperability Conformance Specifications' (ICS) standards for certifying different classes of devices. Such certification marks a major step towards establishing cross-manufacturer, standardized communication.

While certification can confirm the conformity of JDF interfaces, it cannot ensure the optimum interplay of products.

3.4 Interrelation with other print media industry standards

Alongside JDF there is a whole array of industry standards – some recently introduced – that cover the needs of different market segments and processes. These coexist with proprietary standards, and standards from other industries that have an impact on the print media industry. It is essential to the success of JDF that it replaces these standards or is complemented by them. The objective must be to gain an industry standard that enables end-to-end communication within the industry.

3.4.1 JDF and PPF, PJTF, IFRAtrack

JDF is an evolutionary extension of the existing standards Print Production Format (PPF) from the CIP3 Consortium, Portable Job

Ticket Format (PJTF) from Adobe and Job Messaging via IFRAtrack from the IFRA. It will be relatively easy for print service providers who already have some networking based on these formats to convert to JDF interfaces. JDF is built on these standards and includes their functionality. The manufacturers united in the CIP4 Consortium are ensuring that JDF is backwards-compatible.

3.4.2 JDF and Print Talk

Print Talk is an initiative by the print media industry's e-business and order management system suppliers which aims to define the digital data stream between customers and print service providers. In order to automate processes between customers and print service providers, it makes sense to integrate them into the new industry standard JDF. Print Talk and CIP4 have agreed that the two worlds are closely linked and should therefore be mutually compatible. The Print Talk specifications are "built around" JDF and based on cXML, a modified XML standard. Print Talk has initiated major additions to JDF for electronic data exchange.

cXML

Commerce eXtensible Markup Language (cXML) is a variant of XML. This widely-used protocol provides a standard for communication and payment streams between customers and manufacturers. www.cXML.org

3.4.3 JDF and PDF/X3

PDF is an extremely versatile data format that is suitable for use on the Internet, CD-ROMs, in prepress and much more. PDF's wide ranging setting options are a major problem for prepress, since they mean that content data can be delivered that doesn't comply with production requirements. This increases the workload in prepress, only some of which the customer can be invoiced for.

PDF/X3

Portable Document Format/X3 (PDF/X3) is a subset of PDF that restricts PDF's functionality and thereby increases reliability in production.

What is required is a reduction in the number of settings options in order to ensure complete PDFs that cause no problems in production. For this reason, the PDF/X3 standard was established with ISO standard 15930-3. PDF/X3 can be imaged effortlessly with every original level 3 RIP. Preflight checking programs are available to identify and, where necessary, correct any deviations from the standard. Used as a format for content data, PDF/X3 ideally complements the Job Definition Format.

3.4.4 JDF and PPML/VDX

Parallel with the Job Definition Format draft, the Print-on-Demand Initiative (PODi) presented a new XML-based standard for printing variable data in the print media industry.

PODi

The Print-on-Demand Initiative (PODi) is an initiative by prominent manufacturers in the print media industry which aims to develop the market for digital printing. PODi supports the dissemination of the PPML/VDX (Personalized Print Markup Language) standard for digital printing. www.podi.org

PPML/VDX is a digital, interactive container that holds all the parameters for variable printing jobs in object-oriented form. While PDF supplies page-oriented print data, PPML/VDX is used to define the individual objects of a page. This eliminates the need to dispatch the entire contents of pages, as is the case with PDF. Instead, individual objects can be sent to the press and are assigned a name so that they can be accessed at any time from the printer memory. There are overlaps between JDF and PPML/VDX, and the plan is to gradually harmonize these.

3.4.5 JDF and EDIFACT

Print service providers are facing ever louder calls from industrial customers for fully automated job processing solutions. The solutions should enable customers to send orders directly from the ERP system (e.g. from SAP) to the print service provider, without having to print these out and mail or fax them. Internet portals do not provide the ease-of-use and security required for this purpose. One of the interfaces which can be considered for this type

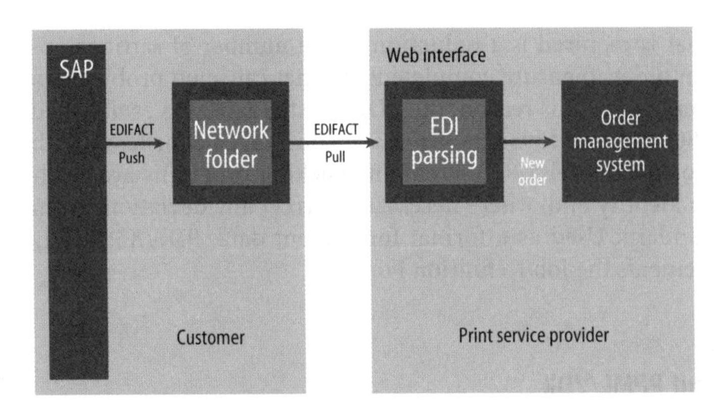

Figure 17
SAP interface to JDF

of data transfer is EDIFACT. Networking with the customer's ERP systems in combination with digital presses enables highly efficient just-in-time production.

This type of data format is converted into a JDF file using a JDF converter. Bi-directional interfaces are possible as part of an export or import function. JDF converters are usually programmed as part of a specific project.

3.5 How secure is JDF?

JDF is based on XML and can be read in plain text by anyone as an ASCII file. If such files are to be exchanged via the intranet or Internet, procedures that ensure a high degree of security must be set in place. Standard procedures are available for this. One possibility is that files are encrypted by the sender and decrypted after they are received or before they are worked on. It is also possible to send the files within Virtual Private Networks (VPNs), which provide secure, tunneled connections over the Internet.

VPN

A Virtual Private Network enables secure communication over the Internet or intranet. The participants perceive each other in the global network as if they were located on a local network. Both encryption procedures and VPNs can be used to securely dispatch JDF all over the world.

4 Networking architectures

Alongside the functionality of the JDF interface, the data transfer between the software applications to be networked is also a decisive factor in the long-term success of any networking project. Data transfer and data management are specified using the networking architecture. The architecture must therefore comply with the following requirements, among others:

- Secure and unambiguous regulation of data transfer and data management
- Data consistency must be maintained,
- Fast data transfer and short reaction times must be guaranteed,
- Unnecessary data transfer must be avoided,
- The entire system must function without error, even when products from different manufacturers are integrated in a workflow,
- The system should include robust mechanisms to make sure that processes are not aborted if problems arise.

Sequential forwarding of files cannot meet these requirements. This is illustrated quite clearly by the following production scenario: In a particular production workflow, a print job is to be split over three different presses, two of which are located within the company and a third at a cooperating print service provider. Finishing is to be carried out with batches overlapping on two different production lines, i.e. finishing work is to start before all the sheets have been printed. Due to deadline problems, the decision to use a second line for finishing is taken at short notice and leads to a revision in planning after printing has started. Given this situation, how can a JDF be transferred sequentially and effectively?

There follows a theoretical account of alternative networking architectures and the most important solutions from different suppliers and initiatives.

4.1 Networking architecture basics

The Job Definition Format is a comprehensive standard for data exchange in the print media industry. However, the JDF specification does not describe what information is to be maintained where, nor who has the right to access and, where necessary, modify it. Basically, the following logical components are provided in a JDF-based networking system:

- *Agents* are tasked with writing a JDF or extending or modifying an existing one,
- *Controllers* route a JDF to the specified location,
- *Devices* form the interface between software applications and machines. Devices interpret the JDF at their specific nodes and control the relevant machines,
- *Machines* are components of the workflow that carry out the processes. They are controlled by a device.

The agents, controllers and devices should not just be able to process JDF though. They should also provide a JMF interface with bidirectional communication mechanism for messaging.

There are two different architectural approaches to JDF-based process integration:

- Decentralized architecture,
- Centralized architecture.

In a *decentralized* networking architecture, a JDF file is sent from one application software to the next. The JDF is saved in the file system or in a professional database by the relevant software application. A decentralized structure in which the JDF is forwarded sequentially is relatively easy to implement and does not require

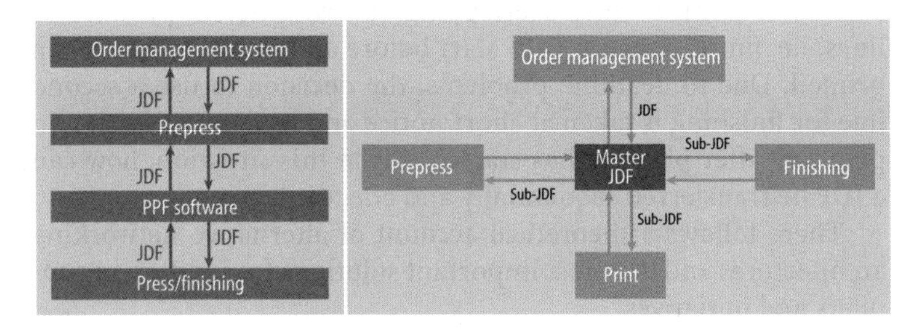

Figure 18
Decentralized vs. centralized architecture

elaborate server architecture. However, the issue of responsibilities does throw up problems. Who is authorized to modify a JDF when and how can it be ensured that the latest JDF is always available to all participants in the workflow? In addition, complex, overlapping processes cannot be reliably mapped.

In a *centralized* architecture, the master JDF or its structure is typically stored in a central database and managed by a server. Databases include various mechanisms for professional data management as standard. These include simultaneous access to one JDF by several participants, reliable processing of semi-completed, aborted transactions, and mechanisms for data replication and backups that also operate during run time. This means that parts of a JDF can be checked out by a software application, changed or augmented, and then checked in again. In addition, the entire JDF, or parts thereof, can be made available in read-only form to other participants. The advantages of storing and managing JDF centrally compared with sequential forwarding of a JDF file from one software application to the next include the following:

- The JDF file is constantly monitored,
- Well-defined checking in and out of sub-JDF,
- Clear access rules define who is allowed to describe specific parts of the JDF tree structure and when,
- Parts of a JDF structure that are currently being modified by other participants can be blocked to all other participants,
- JDF can be checked for consistency and plausibility,
- Better maintenance opportunities.

While a centralized architecture involves more initial work, it does deliver very significant advantages for reliable operation. When implementing a centralized networking architecture, it is important to decide which supplier will provide the centralized server and therefore assume the leading role in the networking architecture.

4.2 Prinect from Heidelberger Druckmaschinen

With Prinect, Heidelberger Druckmaschinen AG offers a series of coordinated software applications that support the end-to-end networking of the entire value adding process. Prinect has its origins in

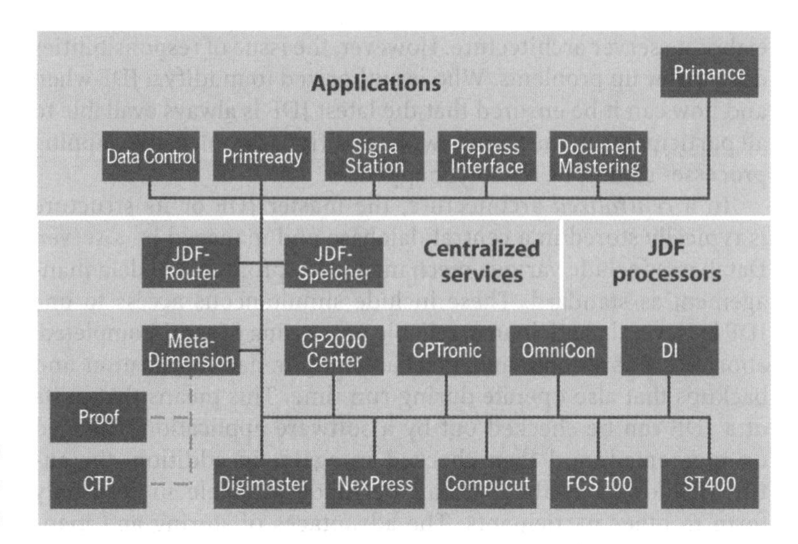

Figure 19 Prinect's future system architecture

a series of software applications that communicate with one another via the Print Production Format and proprietary interfaces. These software applications were developed from scratch and equipped with JDF interfaces. In order to maximize process reliability, flexibility and transparency, Heidelberger Druckmaschinen AG has opted for a centralized networking architecture in which all JDF are stored, spawned and then merged again.

Prinect's networking architecture is a consistent 3-layer architecture which is divided into:

- Applications
- Centralized services,
- JDF processors.

The applications layer includes, among other things, the order management system, the prepress workflow and the production planning system. The JDF is edited, modified and read on this layer. The master JDF is stored and managed on the centralized services layer. Finally, the JDF processors carry out the instructions contained in the JDF and convert these into production processes. After the production processes have been carried out in full or in part, the results are reported back to the centralized services.

4.2.1
Prinect layer: Applications

Heidelberger Druckmaschinen AG has developed the prepress workflow from scratch so that it can be integrated into the new JDF architecture. Other applications have been equipped with a JDF converter, but these are also undergoing new development to ensure optimum integration in the centralized Prinect networking architecture. The goal is to create stable, low-maintenance systems that can be integrated with an absolute minimum of effort. The most important of these are listed below:

- Prinect Prinance (order management system),
- Prinect Data Control (production planning system),
- Prinect Printready System (prepress workflow).

Prinect Prinance is an order management system developed by Alphagraph, who market it jointly with Heidelberger Druckmaschinen AG. This is the module with which Heidelberger Druckmaschinen AG currently has most networking experience. Interfaces have been developed to make job data available for the production process via the Prinect Integration Layer (PIL) integrated in the Prinect Printready System. When the production process is complete, Prinance will in future access the AuditPool elements via the Prinect Integration Layer and make them available for actual costing and statistical evaluation.

Prinect Printready System is a file-based prepress workflow that consistently uses JDF as its internal data format. Using the Prinect Integration Layer, the Prinect Printready System is the interface between the order management system and the centralized services. Instead of using different interfaces between the order management system and the various applications, as was the case before, this minimizes the variety of interfaces and eliminates the need for conversion. Within the Prinect Printready System, too, all the processes are based on the new industry standard. The Prinect Signa Station impositioning software maps the complete JDF structure and the platesetters that are controlled by Prinect MetaDimension (RIP) report their current status (remaining imaging time, plate availability, resolution) to the Prinect Printready System using JMF, thus creating the desired transparency at every workstation. The job data, to which production data has been added in the prepress workflow, is channeled to the centralized services, where it is centrally saved and managed.

Prinect Data Control is the production planning system of Heidelberger Druckmaschinen AG, which combines the data streams from prepress and order management and passes them on to the presses and finishing machines. It is also able to collect real time data from these machines. Alongside the JDF interfaces, Prinect Data Control also features a Host Interface which enables older machines that are not JDF-compatible to be integrated too. Despite the combining function mentioned above, Prinect Data Control should not be confused with a centralized JDF server. Prinect Data Control does not manage the JDF centrally, but rather converts it and passes it on.

4.2.2
Prinect layer: Centralized services

The centralized services form the very core of the Prinect architecture. It is here that the JDF is stored and forwarded to the various JDF processors. They are implemented in the form of a database-supported printroom server which is used to integrate the production resources into the JDF workflow. The centralized services consist of:

- JDF memory,
- JDF router.

In the *JDF memory*, the master JDF is stored, filled in and modified. Parts of the JDF can be checked out by a software application, changed or augmented, and then checked back in again. If errors occur due to incorrect inputs on the application layer or a JDF that is not CIP4-compliant, a consistency check ensures that the workflow is still functioning correctly and error messages are issued to help resolve the transfer problem. The *JDF router* makes sure that the JDF nodes are forwarded to the right JDF processors, read-only access being granted to part of the node, and write access to other parts (e.g. in the case of presettings).

4.2.3
Prinect layer: JDF processors

In future it will be possible to use all of the machines supplied by Heidelberg Druckmaschinen AG as JDF processors. It will also be possible to integrate products from other manufacturers into the workflow as long as they feature certified JDF interfaces. JDF

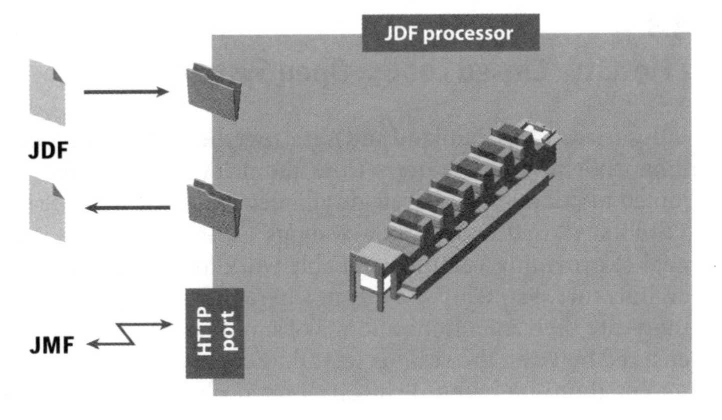

Figure 20
Conversion of Heidelberg products in JDF processors

processors communicate with the centralized services and transmit JMF to the individual software applications via http.

The JDF processors include:

- Prinect CP2000 Center (control console of the Speedmaster),
- Prinect FCS 100 (control console for folders, saddlestitchers).

The *Prinect CP2000 Center* obtains job and machine presetting data from the printroom server as sub-JDF via its ManagementGate and Preset Link modules. This is done by spawning the relevant part of the job for the processor in question from the master JDF, checking it out and dispatching it to the processor. This sub-JDF contains all of the data necessary for the production step. The machine can be preset based on the job and machine presetting data transferred. In addition, a continuously updated electronic job ticket can be accessed at the control console. During this process, messages about the current production status can be returned to the centralized services as JMF. In future, the processes on the press will be documented in the AuditPool elements of the relevant node.

In the case of the *Prinect FCS 100*, the Finishing Communication System, the machine presetting data for the folders and saddlestitchers is captured via the Compufold and Compustitch modules. The Production Data Management returns the AuditPool elements filled in during the production process to the centralized services. Interim messages on the current status are transmitted via JMF.

4.3 PrintCity "Closed Loop... Open Systems"

JDF initiators MAN Roland and Agfa, together with numerous other independent manufacturers from the print media industry, have joined forces to form a strategic alliance under the name PrintCity. With its "Closed Loop...Open Systems", PrintCity has set itself the goal of providing complete, reliable workflows in an open system architecture. PrintCity is a forum where the participants can communicate their intentions and make sure that interface integrity is ensured between the various workflow components.

For the time being, PrintCity's open philosophy does not include a centralized architecture, but the individual manufacturers are able to develop (competing) centralized architectures of their own. Different networking architectures from PrintCity are likely to become established on the market. Some of the software applications that are most fundamental to the PrintCity concept and their interplay in networked production scenarios will be examined below. These include:

- Optimus 2020 (order management system),
- Delano (project management software from Agfa),
- ApogeeX (prepress workflow from Agfa).

Optichrome Computer Systems Ltd., which supplies the order management system 2020, possesses many years' experience networking MAN Roland presses. The existing interfaces have been supplemented by the following JDF/JMF interfaces:

- JDF job description with the key production details,
- JMF messages with up-to-date information for order tracking and actual costing.

The Web-based project management software *Delano* offers publishers and print service providers a communications platform for cross-company collaboration. Delano is available in two versions. *Delano Publish* supports customers in page production with functions made available over the Internet for checking, finalizing, distributing, coordinating and approving the delivered content data. All the activities of staff involved in the project can be tracked. *Delano Production* is the production planning system from Agfa, which can be used to plan and manage all production processes taking into account prepress processes.

- Delano is able to communicate with order management systems via JDF,
- Delano writes a JDF job ticket for Apogee and optionally offers a JDF interface for other PDF-based prepress workflows,
- Delano supplies ResourceLinks for processing print jobs in the correct sequence.

Efficient and comprehensive job ticket functions form the core of the *ApogeeX* prepress workflow from Agfa. The user can, for example, take a JDF from Delano, select job ticket templates from a database and generate new job tickets. Input data accepted by ApogeeX includes JDF job descriptions, JDF job sequence data, JDF runlists, updated JDF job descriptions with reference to the plates produced, and JMF messages from the RIPs that report the status of the platesetter and proofer. The output delivered by the prepress workflow includes JDF job, JDF presetting and JDF production sequence data.

The manufacturers affiliated to PrintCity develop *JDF devices* of varying degrees of complexity for the devices and machines that have to be integrated. PECOM accepts JDF job and production sequence data from the order management system, and preview images and data for machine presettings from prepress. PECOM can be used to check and finalize this JDF data, before using production templates (settings from jobs already printed) to approve it for printing. The machine data is logged and translated into JMF messages. These JMF messages can be used for evaluation purposes in a comprehensive production planning system and made available to the order management system. As a JDF device, PECOM performs process control and presettings for MAN Roland presses. However, on its current platform, PECOM cannot be used as a centralized JDF server.

4.4 Networked Graphic Production (NGP)

Networked Graphic Production (NGP) is based on an initiative from Creo which aims to create a fully-integrated, JDF-based production environment interlinking all players involved in the design and manufacture of printed goods. Apart from Creo, KBA, Komori and Xerox are the main members of this strategic alliance. What is special about NGP is that it has been initiated by the world's leading supplier of prepress solutions, whose market penetration would have given it the potential to successfully market its own centralized server. As with PrintCity, the various manufacturers involved in NGP are independent. Some manufacturers are members of both PrintCity and NGP.

NGP's approach to networking is currently based on a decentralized architecture. Data exchange is ensured using various JDF converters. Prinergy is not able to pass on a complete JDF generated in the order management system to the presses and finishing operations, and neither is an order management system within NGP able to generate a JDF that contains all the information needed for prepress, printing and finishing. Prepress, the printroom and finishing operations receive separate JDFs from the order management system. In prepress, a further JDF is generated that is forwarded to the printroom and finishing.

The NGP initiative gives NGP partners and their customers the opportunity to test their solutions together with other partners. In order to make it easier for partner companies to implement JDF, the NGP suggests a networking architecture consisting of several levels:

- **Level 1:** Unilateral, unidirectional communication of the job ticket from the order management system to production,
- **Level 2:** Bilateral, bi-directional communication between the order management system and production.

One example of a technical link in the context of Networked Graphic Production is the bi-directional exchange of data between Hiflex Print and Prinergy from Creo. A job is automatically created in Prinergy via the standardized JDF interface Synapse Link as soon as the operative has defined it in Hiflex Print's order book. In the other direction, Prinergy provides up-to-date production information such as press busy times, material consumption data, job status and the type of work performed. This ensures that order management, planners and field sales staff remain fully informed about the progress of jobs in prepress and can give accurate answers to customer inquiries. Automatic feedback of operating data and production events from the prepress workflow ensures that the costs actually incurred are billed and incorporated directly into the actual costing process.

4.5
Order management system – a JDF nerve center

Alongside press manufacturers Heidelberger Druckmaschinen and MAN Roland and prepress manufacturers Creo and EFI, ambitious order management system suppliers such as Prism and DIMS! could also take on a key role in the networking workflow. In order for this to happen, it would be essential to configure the relevant planning module as a JDF router, so that the JDF generated in the order management system or Internet portal is forwarded to the software applications or machines in the planned order.

For it to actually be deployed as a centralized JDF router, the order management system must be able with its production planning system to map the entire process structure as well as the product structure. A master JDF is needed to create the necessary transparency and to map more complex workflows. To this end, the machine presettings generated in prepress and all the JDF elements not costed must be managed in the tree structure of the order management system.

It is currently not possible to foresee this concept being implemented in the American market. At the moment, models in which the order management system supplies and receives JDFs via interfaces are in favor. Up to now, no supplier has been willing or able to take on the responsibility for implementing a complete, integrated workflow and make the investments necessary for this.

4.6
EFI

With its takeover of Printcafe in 2003, EFI has become a key player in the spread of JDF in the graphics industry. While Prinect, NGP and PrintCity were developed based on the requirements of offset printing, EFI's offering focuses more on digital printing. The high market penetration achieved by EFI with its controllers and workflows in digital printing, and its acquisition of suppliers operating mainly in commercial printing, opens up an opportunity for practicable, integrated digital and offset workflows. JDF itself could also benefit from this, since it is in the interest of application suppliers to expand the standard for their solutions.

The beginnings of this trend may be seen in the linkup of Velocity OneFlow™, VelocityBalance™ and the Fiery® Controller via JDF. EFI Hagen OA®, the high-end order management system from EFI, is now equipped with JDF interfaces. Nevertheless, the variety of products in the MIS sector, which include EFI Hagen OA, EFI Logic, EFI PSI and EFI PrintSmith, represents a major obstacle to making comprehensive JDF offerings. There is also currently no evidence to suggest that EFI actually wants to develop a centralized JDF server. However, the potential is there. EFI possesses the necessary market penetration and development know-how. EFI's approach is to split up JDF into prepress and press. Finishing is given lowest priority.

Hagen OA with prograph as an integrated production planning tool. This prioritizes JDF, while other integrational steps, such as those between printSmith and Fiery, are based on proprietary formats. This can only be a first, intermediate step.

4.7 PrintNet (ppi Media)

With PrintNet, MAN Roland and ppi Media have conceived an ambitious production planning system with open system architecture for newspaper publishers which they plan to extend to printers and publishers. PrintNet is entirely based on JDF and offers the possibility to integrate third-party products into a comprehensive end-to-end workflow. The PECOM automation and print management system from MAN Roland and the JDF-based PrintNet OM output control system from ppi Media are integral parts of PrintNet. At the heart of PrintNet's architecture is a centralized database which manages the information for all workstations in the workflow. The clients for operating the system are browser-based applications that are not dependent on the operating system.

PrintNet enables the time and place of printing processes to be centrally planned, controlled and statistically evaluated over the entire workflow. PrintNet supports various modules for order processing and tracking, including

- *jobstart*, for creating jobs at a standardized data entry point,
- *jobplan*, for planning production steps with optimized sequence planning,
- *jobperform* for the highly-automated execution of production jobs with flexible workflow control,
- *jobtrack* for tracking jobs over the entire production chain,
- *jobreport* for reports, evaluations and long-term analyses of the production chain.

Production systems are planned automatically based on the print proofing times saved in PECOM, but can also be changed if necessary. External workflow areas at publishers and print service providers can be linked up via PrintNet ports such as officeport (linkup to commercial and administrative systems), newsport (linkup for an integrated editorial workflow), adport (linkup for an integrated advertising department), dtpport (linkup for the creative workflow), storeport (linkup for intelligent material handling) and transport (linkup to the distribution systems).

4.8 Evaluation

Manufacturers also think of the centralized JDF server as the data backbone of print service providers. They therefore place great importance on providing a first-class solution for this data backbone. As with operating systems, this creates the opportunity to develop and sell new, highly-integrated software applications.

With Prinect, Heidelberger Druckmaschinen AG is employing a centralized architecture that administers the master JDF. This server is consistently separated from the application layer so that it can be deployed in any production environment. In the course of developing Prinect, Heidelberg has increasingly moved away from the idea of providing its own solution, and has stepped up its commitment to cooperations with, for example, American order management system supplier EFI for connecting Hagen OA, or with Olive Inc. from Japan. The Prinect Integration Layer serves as an interface to the order management system and PPS. Heidelberg expects a complete JDF from the order management system or PPS, to which further JDF nodes are added in the centralized services.

Both PrintCity and NGP are currently focussed on sequential forwarding of the JDF, a fact which doesn't rule out individual members of these initiatives developing centralized servers. Neither PrintCity nor NGP can therefore be said to have developed a single networking architecture. The endeavors presented thus far must therefore be classified under decentralized architecture, which does not provide reliable and transparent support to complex production processes. On the surface, the offerings from PrintCity and NGP appear similar, but while PrintCity supports the entire JDF standard, NGP uses a subset of JDF it has defined itself with the aim of ensuring the most practicable, easy implementation possible.

The concept of expanding an order management system into a centralized JDF/JMF server, as exemplified by Hiflex's networking concept, is a useful variant offering all the advantages of centralized networking architecture. However, important suppliers on the American market such as DIMS!, Prism and, in particular, EFI with Hagen OA are currently more focussed on providing JDF interfaces. They are not yet prepared to shoulder central responsibility in the JDF workflow. EFI in particular has the opportunity to develop its own offering and create practical solutions to link digital and offset printing that meet the needs of a fully-integrated end-to-end workflow.

The PrintNet concept from ppi Media is pursuing a centralized networking approach that is currently still tailored to newspaper production. This concept delivers all the benefits of a consistent, centralized, database-supported networking architecture which, it is anticipated, will also be used in large-scale commercial printshops in future.

In summary most networking concepts tend toward a centralized architecture. Many of these concepts have only attained an early stage of technical implementation and are still based on hot folder mechanisms. It will still be some time before the individual concepts are mature enough to reliably map all the aspects of the printing process.

5 Benefits

Process integration based on the Job Definition Format helps to eliminate inefficiencies in the production process. This benefit is achieved at the cost of investments in software, hardware and services. In addition, management and the staff undergoing training are tied up to a significant degree during the introductory phase of networking. The benefits and cost of networking must be weighed against one another depending on the order structure and operating structure.

This chapter focuses on the qualitative benefits of the individual networking routes. Empirical statistics from the IRD are quoted to document the scale of the benefits. Following this, a practical example is used to illustrate the benefits and payback period of a networking project.

IRD

The Institut für rationale Unternehmensführung in der Druckindustrie e.V. (IRD) (*Institute for rational management in the print media industry*) is an advisory institute with 700 member companies in the German speaking countries.

5.1 Benefits of e-business

Print service providers and customers alike expect implementing an e-business solution to simplify processes. The benefits must be compelling to both sides. In practice, however, customers will often require print service providers to use e-business solutions as a prerequisite for working with them. Even in this scenario, the print service provider must perform a cost/benefits analysis and develop business models for the new e-business solution.

Receiving inquiries by fax causes inefficiencies in two ways at the print service provider. For one thing, all the data has to be manually entered and edited in the order management system and, for another, incomplete inquiries lead to time-consuming follow-ups or misinterpretations which can adversely affect customer relations. E-business solutions are particularly efficient for very small jobs where the processing time per job using traditional working methods would throw their profitability into question. Internet portals that provide more than just a description of print service providers' range of services are becoming especially widespread in digital printing. Internet portals significantly reduce the processing time, and therefore the process costs, in these types of application. They also enhance customer loyalty (e.g. warehousing, scheduling overviews, image database) and support communication with customers, agencies and print service providers.

E-business solutions enable print service providers to:

- Reduce *process costs* by integrating customers' ordering processes in the order management system,
- Increase *customer loyalty* by integrating with customer-specific processes,
- Extend their *availability*, overcoming time and geographical constraints, thus expanding sales opportunities.

In essence, the customer has similar job processing problems to the print service provider. Some of the process costs involved in issuing and tracking printing orders are significant. Coordination processes can often be dealt with much more quickly and cost-effectively via an Internet portal. Furthermore, the more people are involved in the process, the more likely it is that collaboration via the Internet will pay dividends. With very small jobs there is often no relation between the benefits and the cost of job processing. If the customer's ERP system is not connected with the print service provider's order management system via the Internet portal, the orders first have to be collected in the ERP system, printed out, and then faxed to the print service provider. The provider's order confirmation, delivery note, invoice, etc., are then also entered manually into the ERP system. Changing media in this way is expensive and can lead to problems if production is time-critical.

E-business solutions bring customers the following benefits:

- All of the customer's project members have a complete overview of the project information they need. This includes the product description, the time and place of delivery for order tracking, access to the stock system for requisitioning preprinted materials.

- Accelerated communication thanks to digitization of ordering processes with agencies and print service providers.
- Implementation of the corporate design with templates that are authorized for use throughout the enterprise and can be edited at predefined points in keeping with corporate design guidelines so that they can be adapted for the purpose in question.
- Internet portals typically deliver management information for customers covering inquiries, print orders issued, etc. This represents a source of quick data for checking budgets.

The IRD has conducted an investigation into the deployment of e-procurement which gives some indication of the potential savings involved.

E-procurement

Electronic procurement (e-procurement) refers to the use of Internet-based information and communication technologies to electronically support and optimize the entire procurement cycle of resources and services.

Jobs are sorted into simple, normal and complex, with each category requiring different amounts of time to process. At an hourly rate of pay of Euro 65 and 2.8 quotations for each order actually received, jobs cost between Euro 35 and Euro 218 depending on their complexity. These job costs can be reduced by implementing an e-procurement solution that is integrated with the order management as long as a large enough number of orders is completed via this system to justify the investment and the training, installation and communication costs incurred when the system is introduced.

The more standardized the workflows, the greater the probability that e-business solutions will pay off quickly. Given the great variety of e-business solutions available in the print media industry and cut-throat conditions on the market, it is advisable to choose partners carefully.

Job complexity		Small	Normal	Large
Quotations	Hrs.	0.25	0.37	1.11
	€	16	24	72
Jobs	Hrs.	0.54	0.94	3.35
	€	35	61	218

Table 1
Job costs by order complexity

5.2
The benefits of networked job preparation

The job preparation networking route leads to a standardization of the numbering schemes in order management and production. This enables job, content and production data to be automatically merged at the production control consoles, greatly boosting the reliability of the production process. When job preparation is completely networked, changes are communicated and documented on an ongoing basis.

Networking job preparation should:

- Reduce *job throughput times*,
- Eliminate *inefficiencies* due to duplicated data capture in different software applications,
- Reduce *costs incurred due to errors* by means of standardized designations, up-to-date order information and previews.

A survey by the IRD demonstrates how quickly jobs progress through the individual phases from the inquiry to invoicing (typical values, depending on complexity).

Throughput times for the "Order to confirmation" phase cannot be shortened by networking order management and production. The potential to reduce throughput times can be found primarily in the "Documents to Technology" and "Delivery to invoicing" phases.

The different speeds for processing job tickets can be put down among other things to incorrect, incomplete and frequently changed job tickets. This leads to numerous follow-up questions and high costs incurred due to errors. Investigations by the IRD have discovered that around 90% of all job tickets are modified. Networking job preparation can have a decisive impact here, delivering as it does up-to-date job data quickly to the relevant member of staff's workstation.

Table 2
Typical job throughput times based on IRD data

Job complexity		Small	Normal	Large
Inquiry to quotation	Days	2.2	2.9	4.7
Order to confirmation	Days	1.3	1.6	2
Documents to Technology	Days	0.9	1.2	1.5
Delivery to invoicing	Days	4.2	3.9	5.6
Total	Days	8.6	9.9	13.8

		Not networked	Networked	Difference
Number of follow-ups between Sales and Production	Number/ day	17	7	10
Time per follow-up	Min.	6	6	0
Accumulated time	Hrs./year	382	157	225
Cost per staff member in order management	€/year	24,862	10,237	14,625
Total cost of follow-ups	€/year	124,312	51,188	73,124

Table 3
Benefits of job preparation based on data from the IRD

The sample invoice is based on 5 order management staff with an hourly rate of pay of Euro 65 who work 225 days a year. They have to field between 7 and 17 follow-up questions per staff member per day. Taking an approximate 3 minutes per follow-up and two persons taking part (one asking questions and one answering), the imputed savings can amount to up to Euro 73,000 per year.

While processes do run more smoothly thanks to networked job preparation, the actual benefits are hard to quantify. In the final analysis, the print service provider's staff are relieved of activities that do not add value and can instead take on new tasks that make a greater contribution to the success of the company. The reduction in errors thanks to networking is also difficult to quantify, since usually no definitive values are available that describe the status prior to networking.

5.3 The benefits of networked machine presetting

The presettings produced digitally in order management or prepress are used to set electronically controllable machine components such that operators can initiate the production process as quickly as possible. This networking route has become established with the increasing popularity of CtP, particularly for presetting ink zones on presses. Cutter presetting has also become increasingly popular within specific market segments.

The following benefits are possible by networking machine presetting operations:

- Reduction in *setup times*,
- Reduction in *waste*,

- Improvement in *production reliability and quality*,
- Reassignment of the work to just a few *specialists*.

The decisive benefit of machine presetting, i.e. reducing setup times and waste, is relatively easy to quantify. Thanks to the large installed base, empirical statistics on this are already in place. Up to one proofing operation per job can be saved on a calibrated press. If the ink zone values are not transferred, the printer has to lay every plate on the control console and set the presettings as best he can using the keyboard or stylus. On a fully-automated press a four-color job today can be set up in around 15 to 20 minutes, a process that used to take around 60 to 90 minutes using purely manual methods.[3]

When cutting complex jobs that are not saved on the cutters, staff need considerable time to program all the steps. During this time, no production can take place on the cutter. Using cutting marks from prepress, on the other hand, a sheet can be calculated in minimal time and new jobs prepared during production. The same applies to folders and saddlestitchers, although in this case it is important to consider which presettings can be performed on the machines and whether these justify the networking effort.

Job preparation stations allow trained staff to centrally prepare machines for the production process. The machines and software applications can thus be operated by less highly trained staff without compromising quality.

5.4 The benefits of networked production planning and control

The production planning system schedules the machines and times on a job-specific basis. Simulations can be used to arrive at the best production sequences. Machine scheduling can be changed at short notice. The level of planning detail differs from one company to the next.

The following benefits are possible by networking production planning and control:

- Eliminating *duplicated inputs*,
- Eliminating *follow-up questions* on the production status,
- Reducing the effort required for *planning meetings*,

[3] Prof. Kipphahn (editor): Handbuch der Printmedien, Springer Verlag, 2000, p. 333.

- Increasing the *reliability of production* by ensuring that paper, ink and plates are available when needed at the correct press,
- Increasing the possibilities for the company to control itself through the evaluation of technical information.

In non-networked processes, the planner must re-input the job data (job number, customer, etc.) in his planning tool (e.g. planning chart, Excel). This step is eliminated by networking production planning and control with the order management system. In addition, status messages can be provided to order management, field staff, company management and, if required, customers.

Non-networked production planning and control requires a large number of planning meetings to be held with the departments concerned. These usually takes place in the morning and the plans are then updated several times throughout the day. Usually only the planner, who collects the necessary information from the production staff, has an overview of the current production status. This requires considerable communication input. The IRD has ascertained that planners typically devote less than 30% of their working time to pure planning and controlling activities. Reducing the number of participants in, and the duration of, main meetings and coordination meetings can produce a significant saving which, according to surveys by the IRD, can easily exceed Euro 100,000 per year in midsize operations.

The availability of non-job-specific technical information which is used to document the progress of shifts and machine downtimes is another important benefit. This data delivers valuable information for analyzing non-productive times and disruptions.

5.5 The benefits of networked operating data logging and actual costing

If the company management doesn't want to simply rely on "gut feeling", it needs business management information drawn from the day-to-day operations. As things stand, this information is primarily acquired using daily dockets, but is also increasingly obtained via terminals for logging operating data. Recording, checking and inputting this information from daily dockets is a relatively work-intensive process. The usefulness of this method is limited since the data is frequently not recorded in real time, inputs are missing and entries are posted incorrectly, thereby falsifying the actual result. It is better to evaluate machine data and supplement this with operating data that is not supplied by the machines.

The following benefits are possible by networking operating data logging and actual costing:

- Recording the *actual costs*,
- Eliminating *duplicated inputs* and checks,
- Providing prompt *status messages* and high-quality management information.

While networking the operating data logging process eliminates the need to fill in daily dockets, entries have to be made on machine terminals or operating data logging (ODL) terminals. The real saving therefore lies in the Accounts department, since daily dockets no longer need to be input into the order management system. By evaluating machine data in conjunction with operating data entered by production staff, it is possible to gain a reliable picture of how production resources, times, and consumables are utilized. Random checks can for the most part be dispensed with. The data is automatically transferred to the order management system, where it is available for actual costing and statistical evaluation.

The sample invoice is based on 30 staff in production filling out their daily dockets, with an average hourly rate in order management and production of Euro 65. On average 7 jobs a day are processed, 225 days a year. Staff take 0.4 minutes to fill in each item in the daily docket, and Accounts takes just as long to transfer this information to the order management system. Due to the higher proportion of discrepancies, checking the daily docket takes one minute, twice as long as checking the inputs at the operating data logging terminal. There is no need in the networked system for Accounts to review the inputs at the operating data logging terminals. When using daily dockets around 10 follow-ups are needed per week, which have each been apportioned an estimated duration of 3 minutes. In the sample invoice, the work progress checks could also be reduced from 140 hrs. to 35 hrs. as a result of networking the operating data. This yields potential imputed savings for the sample operation of around Euro 34,000.

Some of the savings that can be made by networking operating data logging and actual costing are quite transparent. This is especially true for the process of transferring information from the daily dockets – an activity that can now be dispensed with. In what measure the elimination of clarification and checking activities actually has an economic impact varies from case to case, since in the phase when networked operating data logging and actual costing is be-

		Daily dockets	ODL/MDL
Writing one daily docket item	Min.	0.4	0.4
Daily docket check by section manager	Min.	1.0	0.5
Input of daily docket item in Order management system by Accounts	Min.	0.4	0
Clarification by Accounts per week	Qty.	10	0
Time per clarification by Accounts	Min.	3.0	3.0
Writing/capturing (staff)	Hrs.	315	315
Check by section manager	Hrs.	112.5	56
Input by Accounts	Hrs.	315	0
Clarifications by Accounts	Hrs.	52	0
Work progress checks	Hrs.	140	35
Total annual cost	€/year	60,743	26,390
Annual saving through operating data logging	€/year		34,353

Table 4
Benefits of operating data logging and actual costing according to figures from the IRD

ing introduced, the values obtained must be checked particularly intensively. Although not factored into the equation here, it will become increasingly important in future to record the additional work occasioned by the customer. In this area, new, networking-enabled software applications can help to document the reasons and the additional work required in these cases.

5.6 The benefits of the networked color workflow

In prepress, color data is edited on the screen and scanner and is then output on the proofer and, finally, printed. If the color workflow from the agency to prepress and all the way up to the press is not standardized, the result has to be coordinated between the customer and printer. The coordination processes on the press cost time and, therefore, money. If the print product is not coordinated, the decision about the coloring of the print job lies in the hands of the printer. Where color fluctuations are significant, jobs either cannot be sold or must be reprinted.

The following benefits are possible by networking the color workflow:

- More *productivity* from the first step to the finished print,
- Increased *reliability* and consistency in the print quality,
- *Shorter setup time* and less waste,

- *True-color proofs*,
- Increased *reliability* vis-à-vis the customer.

The networked color workflow is based on products used for machine presetting, but also includes services and, usually, the use of spectrophotometric instrumentation and control equipment. The networked color workflow enables greater consistency, so that different printers with different levels of experience and skill can achieve consistently high-quality results.

The sample invoice assumes ink zone presetting has already been implemented. The focus is on the improvements that can be achieved by standardizing processes. With 7 jobs per day and 225 working days per year, a paper price of Euro 70 per 1,000 sheets and workcenter costs for the press of Euro 200, a three-minute reduction in setup time and 100 fewer waste sheets per job already yield annual savings in the order of Euro 26,000 per press.

Table 5
The benefits of the standardized color workflow

		Without PCM	With PCM	Difference
Setup time per job	Min.	26	23	3
Setup time per working day	Min.	182	161	21
Waste during setup	Sheet	500	400	100
Waste per working day	Sheet	3,500	2,800	700
Cost of setup time/press	€/year	136,500	120,750	15,750
Cost of material/press	€/year	55,125	44,100	11,025
Total costs	€/year	191,625	164,850	26,775

This information must of course be qualified by the fact that printed paper is often used to setup the press, so that the actual saving is more likely to be around half of the costs presented here. The saving in setup time, too, is in the first instance an imputed value, though this can become relevant in the event of capacity bottlenecks and the expansion of shift operations.

5.7 Payback period of a networking project

Since print service providers often only have limited capital resources, expensive investments must not only provide qualitative benefits, they must also boost the company's financial situation. Companies should therefore always carry out investment analyses. While many are rightly skeptical when it comes to speculating on future events, an investment analysis nevertheless provides a good basis for assessing whether or not investments make economic sense.

In the practical example, the reflow of funds is determined and compared with the costs for investment and ongoing maintenance. The result is used to determine the payback period, which is drawn on as a criterion for the investment decision. A payback period of two to three years is considered a worthwhile investment. If the payback period is longer than this, investments can only be justified by strategic goals such as technological leadership, improved marketing opportunities or ensuring the continued custom of specific customers.

Payback period

The payback period is the length of time it takes for the reflow of funds to cover the total cost of ownership. It is also known as the payoff period.

5.7.1 Basis for calculating the payback period

Companies normally use two key figures in their investment analyses, the Return on Investment and the payback period.

ROI

The Return on Investment (ROI) is a key figure for determining the profitability of investments. The ROI is the product of percentage return on sales (ratio of profits to sales) and capital turnover (ratio of sales to capital invested).

Both key figures are used to make a quantitative statement on the success of an investment. While Return on Investment determines the profits from the investments over a defined period of time, the payback period aims to calculate when the investment costs will be recouped, i.e. paid back. The method of calculating the payback period that is most usual in practice will be used here.

The payback period t is the period of time taken for the discounted, accumulated reflows of funds to recoup the investment costs. Discounted means that the value of money at a defined point in time is taken into account (in periods of inflation, money will be worth more today than it will be in the future). As an alternative to the inflation rate, financing costs can also be used for discounting purposes.

Equation 1: Calculation of payback period

$$0 \leq -C_0 + \frac{C_1}{1+r} + \frac{C_2}{(1+r)^2} + \ldots + \frac{C_t}{(1+r)^t}$$

C_0 = Investment
C_1 = Reflow of funds, first payment period
C_2 = Reflow of funds, second payment period
C_t = Reflow of funds, final payment period
r = Interest rate for financing costs

The payback depends on the monthly reflow of funds, i.e. the assumed earnings per month less expenditure – or, alternatively, the savings that can be attributed to the investment. The payback period does not provide information about the period in which a reflow of funds will occur, nor about the value the investment will add (since the period after the investment has been paid back is not considered). The investment analysis carried out before the decision was made to invest should be compared with actual values for the purposes of project auditing.

In this context, a distinction must be made between payment-related and imputed savings. Payment-related savings have a financial impact in the form of reflows of funds, while imputed savings are not directly connected with a reflow of funds. Imputed savings only have a payment-related impact if staff or production capacities are cut back.

Payment-related costs

Payment-related costs are costs that have a financial impact. They are contrasted with imputed costs such as owner's salary, cost of capital, etc.

5.7.2 Practical example

The print service provider in this example works with 38 staff in a two-shift operation on four mid- and small-format presses. These presses are utilized 15.5 hours per day. The company serves around 300 customers, processes approx. 2,500 jobs a year and writes 5,500 quotations. The range of products breaks down into 30% standard products such as business stationery and advertising printing, 50% complex products such as brochures and calendars, and 20% periodicals such as journals and magazines.

The company, which is run by the owner, has already gained some early experience of networking concepts. Initial attempts were undertaken in 1998, but failed due to the lack of integration readiness and weak functionality of the order management system. This led to the company changing suppliers in 2000. Equipped with a completely revamped machine park, the three-stage operation then successively introduced the presetting, job preparation, production planning and control, color management, and operating data logging and actual costing networking routes. The print service provider opted for Prinect, the networking offering of Heidelberger Druckmaschinen AG.

The scope and content of the networking project were defined in a two-day workshop. A decision was made not to network finishing operations to begin with. All attention was focussed instead on the printroom, the sector that yields a considerable portion of the added value. The project was launched with the introduction of ink zone presetting and the order management system. Once the order management staff had become familiar with the order management system, initial efforts concentrated on standardizing the color workflow using spectrophotometric measuring equipment and setting up the job preparation and production planning and control networking routes. After six months, the processes were sufficiently well established for networked actual costing to be introduced.

The introduction of operating data logging and actual costing was prepared in detailed discussions in order to achieve greatest possible acceptance among the workforce. Although workers were initially skeptical about complete time recording, many staff were surprised to learn how much time is spent on which activities, particularly ancillary activities.

The benefits of networking were estimated with the help of an ROI tool. This was done by comparing the status after the networking project had been completed with the unsatisfactory, non-functioning partial networking that had existed prior to this. Adequate information on the non-networked status was no longer available. But this does not detract from the usefulness of the practical example, since the poor networking at the investigated plant led to the adoption of working practices that came very close to those of a non-networked status.

5.7.3 Quantitative benefits of networking

The print service provider in the example was able to reduce its order management staff from 4 to 2 in the course of the networking project. Half a post was not refilled, and one further person has been assigned to alternative duties internally. These staff cuts in order management were made possible by more efficient job preparation and operating data logging workflows.

Networking order management provides the print service provider with an overall saving of 16.7%. This breaks down into a 21.0% saving with standard products, 13.2% with complex products and 11.0% with periodicals. This saving corresponds to around 1,200 working hours, and is based on far better coordination than was the case previously. This is mirrored both in a reduction in the number of meetings and a dramatic decline in follow-up queries. The improved flow of information cut the number of follow-up queries between order management staff and production from a total of 40 per day to 10 per day.

The automatic transfer of machine and operating data into the order management system eliminates around 330 hours of work inputting information. In addition, daily docket checks by the production manager and the associated follow-up queries by order management were halved. The reduction in checking work was quantified at approx. 50 hours.

The complete recording of all activities identified Euro 95,000 worth of additional services that had been rendered, though only Euro 45,000 of this could be passed on to customers. The rest was classified as goodwill.

The company invested around Euro 180,000 in Image Control to network its color workflow. This investment resulted in significant savings.

Press setup times were cut from 20 – 25 minutes to 10 – 15 minutes. With around 12 setup procedures per day and press, 100 less waste sheets per setup procedure, workcenter costs of Euro 150 per press, 225 working days per year and an average paper price of Euro 35 per 1,000 sheets, this resulted in a total saving of Euro 76,950 per press. Reduced setup times saved Euro 67,500 of this and reduced paper consumption Euro 9,450.

5.7.4 Evaluation of the investment decision

The payback period is determined by considering two areas separately. These are, on the one hand, networking of job preparation (JP), production planning and control (PPC) and operating data logging (ODL) and actual costing – grouped under the term operational data networking – and, on the other, networking of machine presetting and color workflow.

Year		2001	2002	2003	2004
Investment: JP, PPC, ODL	€	−120,000			
Software maintenance	€	−8,000	−8,000	−8,000	−8,000
Reflow of funds Staff costs (1 post)	€		+60,000	+60,000	+60,000
Net monetary value after reflow of funds (r = 10%)	€	−128,000	−88,800	−29,680	+19,352

Table 6
Payback period for operating data networking

The initial investments and ongoing expenditure on software maintenance contracts are set off against a reflow of funds deriving from savings in staff costs. Imputed interest of 10% is assumed. A new order management system is needed and has a major impact on the level of investment required. The networking implemented thus far does not support the recording of additional services, which therefore have no influence on the payback period.

If the costs for the order management system are included in the payback period analysis, the operating data networking solution achieves a positive net monetary value in the fourth year. Without the order management system, the investment would have paid for itself as early as the second year.

When it comes to the networking of machine presetting and the color workflow, the initial investments in Prinect Image Control, services and ongoing expenditure for maintenance contracts are set against reflows of funds from savings in setup times and reduced paper consumption.

In theory these networking routes will yield a positive net monetary value after two years. In practice, however, this payback will only have a payment-related effect if the company is operating at its capacity limits. In the example, the savings in setup time eliminates the need for a third shift at peak times.

Year		2001	2002	2003	2004
Investment in color workflow	€	−180,000			
Reflow of funds, setup time per press	€		+67,500		
Reflow of funds, waste per press	€		+9. 450		
Net monetary value after reflow of funds for four presses(r = 10%)	€	−180. 000	+109. 800		

Table 7
Payback period for networking the color workflow

In summary, while they can be directly attributed to the networking, the savings described here are not sufficient to provide clear justification for the entire networking. When making purchasing decisions, however, qualitative gains resulting in improved workflows, reduced proneness to error, increased customer satisfaction and more efficient operational management must also be considered. It is difficult to record qualitative criteria in reliable, black and white figures, but these important arguments should be carefully considered when deciding on investments.

6 Procedure when implementing process integration

Process integration has a major impact on the print service provider's processes. Incorrect decisions and organizational shortcomings when preparing for networking can have a lasting disruptive effect on the production process. Therefore careful, prudent project management is the order of the day when planning and implementing networking projects.

It makes sense to start the project by making sure the prerequisites for networking are in place. Obstacles faced by networking projects can include in-house feasibility concerns, for example objections by the works council, and the lack of a systematic structure in the existing workflows. But an unsuited machine park can also send networking costs rocketing to unjustifiable levels. If the prerequisites for a successful networking project are satisfied, the step-by-step implementation can be started. After the networking project has been completed, the newly-introduced processes must be monitored on an ongoing basis.

6.1 Prerequisites for networking

Companies should make sure that the organizational and technical prerequisites for networking are in place before deciding whether or not to invest in networking projects. It may sometimes be advisable to put off the actual networking investment and first satisfy the prerequisites necessary for networking.

6.1.1 Technical prerequisites

The technical prerequisites basically relate to the networking capability of the order management system, the prepress workflow and the machines.

The networking capability of order management systems varies greatly. Interfaces that support machine data logging for actual costing do not yet exist in many cases, and there is as yet no experience of transferring process plans from order management to prepress. Within the foreseeable future only a few CIP4 Consortium-certified order management systems that support the full scope of JDF will be available on the market.

In prepress, the capability to network with the order management system will initially be established for the new workflows that are either JDF-based or equipped with JDF interfaces. The creative portion of prepress, which often makes up a key element of the work, represents a problem in that it has not yet been possible to integrate layout and image processing programs in the JDF architecture. A few software houses are offering database-supported solutions that record the time spent processing documents.

Readying installed presses for networking so that they can exchange state-of-the-art data formats with the order management system can be a costly business. Costs can be so high as to render networking unviable. Older presses can be integrated into a networking architecture using operating data logging terminals. A few suppliers of order management systems offer external sensor systems that record the number of sheets printed or folded for cases such as these.

In all of these cases, checks must be made as to whether an economically justifiable upgrade is available that covers the desired scope of functions, or whether it is best to invest in a new solution that offers better prospects for the future.

6.1.2 Organizational prerequisites

Process integration demands a high proportion of costed jobs, since this is the only way the information required in production can be provided. Before a networking project starts, therefore, processes should be established in which costing takes place throughout. Changes should be recorded by order management or by the relevant member of the production staff, so that continually updated job tickets are available to the latter and useful target/actual comparisons can be made. Daily dockets should be as entrenched a part of operations as actual costing and statistical evaluation of the figures produced. Additional work occasioned by customers should be recorded.

Company management plays a key role in networking projects. It must possess a clear entrepreneurial vision and be able to motivate staff to get behind the changes. Networking brings with it major organizational and technical changes that can drag on for a long time. These must be accompanied by appropriate communication and training measures.

6.2 Step-by-step implementation

Bringing networking projects to a successful conclusion requires consistent project planning and guidance. Networking projects are often so large that the first step should be to carry out a thorough analysis of the starting situation. Therefore, every networking project starts with a requirements analysis, which is often augmented by a process costs analysis. These phases come to an end when the functional specification is adopted. A suitable partner can be selected on the strength of this functional specification. Only then should the organizational measures be carried out and the technical concept implemented. The final phase involves ongoing monitoring of processes based on key figures obtained in the analysis phase.

6.2.1 Requirements analysis

There are many types of requirements analysis, depending on the print service provider's clarification requirements. If the print service provider knows exactly what investments it can make to eliminate bottlenecks and cut process costs, or has already selected a particular supplier, it can have the supplier draw up an inventory, which leads to a functional specification and, finally, a quotation.

Companies who fail to carry out a detailed requirements analysis run the risk of overlooking key aspects of networking and making misplaced investments. Against this backdrop print service providers are recommended to conduct a structured, open-minded staff survey aimed at identifying potential improvements in existing workflows, pinpointing organizational and technical shortcomings and defining the focus for investments. The requirements analysis must cover all of the print service provider's processes and help to define the correct dimensioning required for networking, including the workstations needed. The end product of the requirements analysis is a functional specification that clearly sets out the networking project's functional scope and implementation steps. The specification should include detailed descriptions of the organizational and technical measures necessary. The acceptance criteria for the networking project should also be listed.

Although it takes time to perform a detailed requirements analysis, experience shows that the effort spent on producing a clear, detailed specification is normally recouped many times over when the project is implemented. The print service provider in question must decide whether to carry out the requirements analysis in-house or to enlist external help. Print service providers who opt to carry out the requirements analysis in cooperation with one or more suppliers run the risk of the networking concept being tailored to the supplier's product portfolio without alternative solutions being considered. In any case, the member of staff entrusted with implementing the project should be relieved of his/her daily tasks as much as possible, since conducting a requirements analysis is a time-consuming business. An external consultant can create a manufacturer-neutral concept and accompany it to implementation. Such concepts form a good basis for conducting well-informed negotiations with multiple suppliers and deciding between different networking concepts.

6.2.2 Process costs analysis

The process costs analysis uses a typical product mix to reproduce the key business processes of a print service provider from the quotation phase to delivery. This is done by surveying the individual activities that are performed by staff in processing an order. The current situation is recorded, weak points and inefficiencies are identified and, where necessary, suggestions for improvement submitted. The potential improvements are quantified in monetary terms using benchmarks.

Benchmarking

Benchmarking is a strategic controlling tool used to compare value adding processes, management practices, products or services between companies or business units within a company with the aim of discovering performance shortcomings.

Process costs analyses can require enormous amounts of work, since they penetrate down to the level of individual activities. As a result, they are often not performed in practice, despite delivering important information about a company's reorganization requirements and assessing its success.

In the process costs analysis, indirect costs are assigned to cost units using the cause and effect principle. The fact that the process costs analysis records all the indirect costs of the indirect functional areas makes it a full cost system.

Full cost system

While a full cost system allocates all costs to the relevant allocation bases (e.g. cost centers, cost units), direct costing only takes into account the costs that are relevant for the specific purposes of the cost system.

In the process costs analysis, the process chains are analyzed in terms of organizational and technical workflows, direct costs and indirect costs that can be broken down to specific processes or jobs. The following distinct steps are involved:

1. The cost driving processes in procurement, production, administration and/or sales are identified and subdivided into individual activities. In every cost sector, a distinction is made between output volume-related and non-volume-related processes. In the case of volume-related processes, a dependency exists between the volume of work involved in a process and the output volume of the cost center, e.g. the runs on the "press" cost center. In the case of non-volume-related processes, e.g. job preparation work, this dependency does not exist.
2. Individual activities are grouped into operations that lead to a specific work result in a cost center. For example, it doesn't make much sense to separately analyze every single process within the prepress workflow, such as preflight, trapping and color management. These activities are grouped into the "form production" operation.
3. The cost determinants (cost drivers) are established and allocation bases that reflect the scope of operations and cost levels are defined. Part of the costs can be assigned to the process using the cause and effect principle (output volume-related), while another part cannot be assigned, however (non-volume-related).
4. Indirect costs occasioned by an operation and, therefore, by a cost driver, are grouped into cost pools.
5. One or more indirect cost allocation rates are defined for the grouped operations. The used-capacity principle or the principle of average is used for these non-volume-related costs. The

indirect costs are allocated to the cost units in accordance with the selected allocation base using percentage indirect cost allocation rates.

A process costs analysis produces information on what the individual jobs cost depending on their complexity. Job costs are determined using job order costing, and can then be used to quantify inefficiencies and gage the cost-effectiveness of the product mix. Among other things, process costs analysis can be used to clarify which product groups can be produced with costs being covered. Company management can decide whether it wants to continue offering product groups that do not cover costs in order to achieve an additional contribution margin, whether it can cut indirect costs or should make a de-investment.

6.2.3 Choosing a suitable partner

Choosing a suitable partner is a key task in any networking project, since in most cases it means indirectly choosing a networking architecture. In practice there are a number of constraints, which may for example have their roots in the existing machine park. Nevertheless, companies usually have two to three alternatives to consider. Selecting the best among these is not always easy, since at the time the decision is made companies often do not possess the knowledge needed to determine which concept will provide the best long-term solution.

It is advisable to agree that a partner take on a quasi general contractor role. The general contractor is responsible for drawing up a functional specification together with the other suppliers. He must also ensure that installation and service proceed smoothly and takes on full responsibility for the project during the entire life cycle.

The following aspects should be considered when choosing partners. The weight attached to each of these aspects can differ depending on the company and the project circumstances:

- The networking architecture offered by the supplier is a key criterion when choosing a partner. A centralized architecture that delivers efficient, consistent handling of JDF over the entire production process is clearly preferable.

- The partner should possess both good technical and commercial expertise. While technical expertise helps when implementing networking in the production workflow, commercial expertise is needed to provide advice to customers in advance and set up an informative actual-costing system.

- The partner should have experience of previous networking projects. Adequate experience can help reduce the volume of problematic details thrown up in such projects. For example, the manufacturer-specific private sections are often a hurdle to achieving end-to-end data transfer.
- Networking problems become apparent at the interfaces. A centralized service for all the workflow components can help to quickly get problems under control.
- Since networking can also be used to optimize data throughput in the network, make better use of resources and consolidate the server landscape, it is also helpful to have a supplier experienced in servers, networks and storage solutions.

6.2.4 Organizational measures

A networking project provides companies with an opportunity to critically review and improve their organizational structures and all the processes they contain. It doesn't make sense to network a plant without also considering changes to the structural and operational organizations. Often, reorganizational measures embarked on in preparation for a networking project in themselves achieve significant savings. Organizational changes should always be put in place before the technology is introduced. This ensures that the new technology can be immediately applied to the improved structures (removal of interfaces, integration of work steps), thus eliminating the complex task of adapting it to the improved structures at a later stage.

The focus must be on integrating processes and abolishing interfaces. There is no point in networking an operation at a technical level based on an unstructured operational organization. It takes a great deal of effort to automate unstructured workflows. Processes from job preparation, production planning and control to actual costing should therefore have a clear structure. This will increase the amount of responsibility placed on order management, since the latter has a direct impact on production in networked operations. Other areas will be relieved of non-value-adding activities.

Implementing networking projects has a dramatic impact on workflows. Order management gains a far greater influence on production than is the case in non-networked operations, a fact that calls for new arrangements regulating the areas of responsibility. The following variants can be considered, depending on the scope of networking and staff's level of expertise:

1. Order management as sole manager: Requires that order management possesses a high degree of production expertise. In this scenario, order management also takes over production planning and control. Usually requires technical support, for example from prepress staff. There is currently a trend among print service providers to group order management together with prepress both organizationally and spatially.
2. Order management and technical job preparation: Classic organization at print service providers with increased planning requirements. Technical job preparation plans production and ensures that print products are delivered on time. Process integration relieves the workload on planners, freeing them up to take over other duties, for example in order management.
3. Order management as an interdisciplinary team involving office-based staff, job preparation, prepress and postpress: Extremely labor intensive way of coordinating production processes, since a lot of time must be set aside for planning sessions. Process integration should help reduce the large amounts of communication this organizational form requires.
4. Order management plus distributed knowledge base: The responsibility lies with order management. The JDF is augmented at the individual workstations, and a centralized JDF server ensures the transparency necessary when changes are made. This variant requires a well-defined distribution of tasks. It only works with highly standardized processes.

Basically, companies should aim for a clear structure with integrated processes. In principle, data should only be input once and should be made available to relevant staff on a decentralized basis by means of transparent mechanisms. This is the only way to achieve well-maintained data that remains consistent over the long term. Clear arrangements must also be put in place governing:

- who,
- is allowed to change what data,
- when,
- and under what conditions.

Laying down clear guidelines on these issues is paramount to ensuring that data remains consistent and coordinated.

6.2.5 Technical implementation of networking

After the functional specification has been drawn up and the necessary organizational measures have been put in place, the technical implementation can be dealt with. In this phase it is important to subdivide the project as a whole into individual, self-contained phases that can each run in real time operation while the next phase is being implemented and tested. This is best achieved using a concept with individual networking routes that can be introduced step-by-step. Each phase is completed when an acceptance test has taken place in line with the criteria defined in the function specification.

Real-life experience has shown that that it takes around 9-12

Phase before order awarded	Duration
Requirements analysis	Approx. 1 month
Process costs analysis	Approx. 2 months
Choosing a suitable partner	Approx. 1 month
Organizational measures	Approx. 3 months

Technical implementation of networking	Duration
Introduce quotation and order management software	Approx. 3-6 months
Install network in printroom and finishing area	Approx. 1 week
Retrofit network capability to presses and finishing machines	Approx. 2 weeks
Introduce e-business	Approx. 2 months
Job preparation	Approx. 2 weeks
Machine presetting	Approx. 1 month
Production planning and control	Approx. 1 month
Color workflow	Approx. 2 months
Actual costing	Approx. 2 months

Real-time operation	Duration
	Approx. 2 months

Table 8
Example of networking process steps

months to implement complete networking projects. Table 8 gives an approximate idea of the duration of each phase. Some of the tasks in the individual phases can be worked on in parallel. For example, while the new order management system is being introduced and run in parallel with the old system to ensure data consistency, machine presetting can be taking place at the same time. E-business can also be launched separately from the introduction of other software applications. Following a set order only becomes important when various software applications are being integrated. The speed at which complete networking can be implemented is determined by the ability of staff at the print service provider to familiarize themselves with new software applications and learn new processes.

6.3 Verifying success and ongoing optimization

What benefits a company actually derives from networking depends largely on the commitment of its management. Networking technology provides the infrastructure for making the necessary figures available efficiently and quickly. However, while a networking system may be fully operational on a technical level, it can only be transformed into a successful networking project if conclusions are drawn from the analyses and evaluations and put into action. The accuracy of controlling is completely dependent on the care put into recording the data. It is therefore vital to ensure that the actual-costing process is reliable and suitably structured.

The following sections look at the duties of the controlling department and the statistics that are important for management. This is intended to clarify what data can be made available through networking and what it can be used for.

6.3.1 Duties of the controlling department

One of the major benefits of networking is its ability to produce high-quality actual data for management. This data must be summarized and interpreted by the controlling department. Detailed statistics and an informative system supplying KPIs can be used on an ongoing basis to identify which processes are not running as expected and which product categories are not profitable. Management must then decide what organizational or technical measures are needed to achieve the desired improvement.

The controlling department focuses on customer benefits and works with benchmarks, actual values and planned values. The four key criteria of costs, quality, service and time are especially important for print service providers' competitiveness, and these must be measured. In practice, however, the focus is often on economic efficiency overviews.

6.3.2 Key management statistics

The statistics listed here are examples of figures that can be used for management purposes. They give a rough overview of the operating data which normally has to be evaluated. These statistics must be adapted and augmented to meet the needs of the business in question. The following list therefore makes no claims to completeness.

- The *overview of results* depicts the results in job-specific form. If a job is marked as completed when the daily dockets are entered, the job's result, contribution margin and percentage return on sales appear in the order management system's statistics. This list can be used for checking and evaluating actual costing of individual jobs.
- The *customer analysis* depicts sales data and conversion ratio, the ratio of quotations to actual orders for each customer. A comparison with the previous year's values provides an indication of current and future customer loyalty.
- Product groups' sales, contribution margins and costs can be found in the *product group analysis*. The product group analysis reveals which products are yielding profits and which are unprofitable. Thorough recording of process disruptions often shows that customer specifications or product tolerances for certain product groups have caused overly strict quality criteria to be defined for production. Such product groups, or even customers, do not deliver profits.
- The *cost center statistics* provide and productive and non-productive times for every cost center and also determine key business management indicators such as the capacity utilization rate and non-productive time ratio. Given that fixed costs constitute up to 80% of printshops' total costs, companies should always keep an eye on the capacity utilization rate, which reflects the number of value-adding production hours. The cost center

statistics also give a clear picture of machine downtimes due to faults. These can be used to introduce appropriate measures for ongoing maintenance of machines or to prepare for the next investment.

- When recording *faulty and duplicated work*, a distinction is made between process fluctuations and process disruptions. Process fluctuations are randomly occurring fluctuations linked to the production process (e.g. errors in the offset plate copy process). The costs incurred and time lost due to errors in this process reduce the cost center's capacity utilization level (increase the non-productive times and, with this, the hourly rate). These costs cannot be allocated to any job. Process disruptions are organizational disruptions in the planned sequence of jobs that can be attributed to a specific circumstance and do not occur randomly. The costs are related to cost units and reduce the profit of each job. Incorrect or duplicated work leads to expensive disruptions in the production workflow and can be reduced using suitable countermeasures. An analysis of errors must determine whether the customer can be held liable for the economic losses incurred, whether specific production guidelines or designated quality checks were not adhered to, and whether minor investments would help reduce the incorrect and duplicated work.
- The *additional work statistics* record services provided in response to the customer delivering faulty files, requesting changes too late, etc. This includes full costs for working time, materials and machine downtime. Company management decides whether the customer can be invoiced for these costs.

7 Closing remarks

In the past, print service providers were able to achieve significant increases in productivity by investing in more and more powerful machines and automating individual production processes. However, the potential boosts in technical efficiency that can be achieved in this way are now exhausted. The next impetus for development is being provided by the integration of the entire value added chain including external partners.

The JDF industry standard provides the technology needed for comprehensive process integration in the print media industry. For the first time, JDF enables manufacturers to integrate different production environments based on a system of interfaces. Manufacturers and initiatives have developed different networking architectures to this end.

When purchasing new machinery and software in future, greater attention will have to be paid to their networking capability. For companies aiming to completely network their operations, a centralized networking architecture is preferable, since this is the only way to ensure a consistent, transparent transfer of data.

In complete networking projects in particular, it can make sense to select very few from among the many suppliers on the market in order to avoid potential interface problems.

However, the best technology is to no avail if the required organizational measures are not consistently implemented and supplier support in the form of consulting services, customer services and the provision of comprehensive documentation is lacking.

The technologies and products for process integration are out there - it's now up to corporate decision-makers to recognize and realize the opportunities they provide.

Annex

The following checklists are intended to help the reader identify process inefficiencies and devise a suitable networking concept in preparation for a networking project. The checklists make no claim to completeness.

Checklist for reviewing process inefficiencies

1. Are the business and production processes documented?

Process documentation is a useful basis for checking that the organizational prerequisites for process integration exist. If standardized processes are not established, the organizational prerequisites for networking should be put in place first.

2. Are orders from field staff correctly costed?

When orders are costed by field staff who have no access to centralized databases, the calculations may be incorrect because the cost factors are not properly taken into account.

3. Are inquiries, orders or delivery notes produced manually using word processing and collated data?

Methods like these allow no data consistency. They are prone to errors, time is lost through work being duplicated several times, and the safe storage and forwarding of information is not always ensured.

4. Is order management data administered locally?

When information such as data for generating inquiries, technical orders and the choice of suppliers for planned outsourcing is maintained locally, not all order management staff are made aware of changes. Maintaining data locally is basically fraught with problems.

5. What percentage of orders are costed in the order management system?

Networking order management only makes sense if the proportion of costed jobs is very high. This is the only way to make sure that the necessary data is available in production and meaningful actual costing and statistical evaluation can be carried out.

6. Are jobs processed in production without being costed in detail, or does the scheduling change vis-à-vis that defined in order management?

These are all good reasons to consider a production planning system. The production planning system becomes an absolute must when implementing networking projects.

7. Does order management receive quick, complete feedback from all production areas?

Order management needs detailed information to keep customers informed and for invoicing them. Changes undertaken at short notice in production due to changed customer specifications or bottlenecks occurring must be sufficiently documented on the job ticket, otherwise the procedures will not have the transparency needed when invoicing the jobs.

8. Is operating data logged only on daily dockets, or not at all?

Daily dockets are often not promptly filled in, and are of limited reliability. The manual transfer of daily dockets for actual costing represents an unnecessary operation.

Checklist for devising a networking concept

1. What payback period are you aiming for?

The shorter the payback period, the lower the investment risk. You should normally aim for 2–3 years, depending on what in-house guidelines say.

2. How is investment controlling to be assured?

Goals and measurement variables for assessment purposes should be defined in advance, because it is often difficult to assess the success of an investment in networking.

3. Who should carry out the requirements and process costs analysis?

The analysis can be carried out by an external consultant or an in-house project team. If using an in-house project team, make sure that team members are relieved of their day-to-day duties so that the project can be processed promptly.

4. Is professional project management deployed?

Professional project management should always be deployed to accompany and safeguard networking projects.

5. Is there a functional specification for the networking project?

A detailed functional specification is needed to ensure transparency about the functional scope of a networking project.

6. What experience do project partners and suppliers possess?

Given the complexity of the subject, comprehensive experience of networking projects is essential. Networking projects often founder on the details.

7. Which networking routes bring the greatest benefits?

Identifying the benefits helps companies make the best use of investment capital. It is advisable to let economic sense prevail over technical feasibility when specifying the implementation.

8. What cost centers should be networked first?

Networking can be introduced gradually. The starting point is usually the area which adds most value, the printroom. Retrofitting older presses can be expensive and, in some cases, can be circumvented using external operating data logging terminals.

9. What scope of functions must the order management system fulfill?

The order management system administers all jobs. The functional requirements must be met without compromise.

10. What IT infrastructure is the networking based on?

Print service providers should select an IT infrastructure so that the IT department can operate and troubleshoot the installed systems (hardware, operating system and network) at short notice without outside help.

11. Which system maintains the master data? Which system ensures the consistency of the JDF?

Administering the JDF centrally makes for a reliable, powerful architecture. The supplier of the JDF server takes on the central integration role.

12. When is a master JDF required, and when is a sub-JDF required?

In non-centralized concepts, the different suppliers must coordinate in great detail who administers the master JDF when and who receives a sub-JDF for carrying out individual tasks when.

13. What data is transferred in what format?

Companies should aim for a completely JDF/JMF-based architecture. In practice, however, older systems must also be integrated via proprietary interfaces. The functional scope and transfer mechanisms for doing this must be clarified.

14. To what extent is JMF implemented, which processes are not supported in JMF?

For transferring up-to-date messages, JMF should be implemented via a bi-directional protocol.

15. Have the JDF/JMF interfaces been certified?

CIP4 and GATF have signed a cooperation agreement to develop certification procedures with the aim of increasing the reliability of interfaces in the future. Until then, company-specific certifications can offer some measure of reliability.

16. To what extent are JDF private sections used?

It is important that private sections are only actually used for data not defined within the standard scope of JDF.

17. How should system acceptance testing be carried out?

A precise system acceptance test must be carried out following installation to make sure that all the system components and data transfer work faultlessly. The appropriate tests and acceptance criteria should be agreed by the supplier and customer in advance.

18. Is comprehensive service in place for all the networked products?

An efficient method of solving interface problems must be in place. As it is not exactly helpful when suppliers start blaming each other, defining a type of general contractor role can create clear lines of responsibility.

Organizations

BUW/DMT With its print and media technology courses (*Druck- und Medientechnologie – DMT*), the Bergische Universität Wuppertal (BUW) offers college-level courses of study and research on networking in the print media industry.

Further information can be found at www.dmt.uni-wuppertal.de

CIP3 The International Cooperation for Integration of Prepress, Press and Postpress (CIP3) stewards the Print Production Format (PPF), an industry standard developed principally by the Fraunhofer Institute for Computer Graphics (IGD) in 1993.

Further information can be found at www.cip4.org

CIP4 The International Cooperation for Integration of Processes in Prepress, Press and Postpress (CIP4) is the successor organization of CIP3, and has over 200 member companies. CIP4 is developing the Job Definition Format and is working to promote its adoption in the print media industry.

Further information can be found at www.cip4.org

EUPRIMA The European Print Managementsystem Association (EUPRIMA) is a consortium of order management system suppliers with the common goal of developing practical solutions for the process integration of print service providers, customers, suppliers and subsuppliers based on the Job Definition Format.

Further information can be found at www.euprima.org

GATF The Graphic Arts Technical Foundation is a non-profit organization that supplies the print media industry with technical information and services. GATF partners CIP4 in the certification of JDF interfaces.

Further information can be found at www.gain.net

IFRA IFRA's origins go back to the "INCA-FIEJ Research Association", where "INCA" stands for "International Newspaper Colour Association" and "FIEJ" for "Fédération Int. des Editeurs de Journaux".

Further information can be found at www.ifra.com

IRD The Institut für rationale Unternehmensführung in der Druckindustrie e.V. (IRD) (*Institute for rational management in the print industry*) is an advisory institute for operations in the printing and paper processing industries with approx. 700 member companies in the German speaking regions.

Further information can be found at www.ird-online.de

ISO

The International Organization for Standardization (ISO) is an international organization dealing with the development, adoption and documentation of international standards.

Further information can be found at www.iso.com

NGP

Initiated by Creo, Networked Graphic Production (NGP) is a strategic initiative with the goal of creating practical JDF implementations. NGP gives member companies and their customers the opportunity to test integration solutions.

Further information can be found at www.ngppartners.org

PODi

The Print-on-Demand Initiative (PODi) is an initiative by prominent manufacturers in the print media industry which aims to develop the market for digital printing. PODi supports the creation of standards for digital printing.

Further information can be found at www.podi.org

PrintCity

Following the initiative of MAN Roland, several companies from the print media industry banded together to form a strategic alliance called PrintCity. PrintCity has the goal of providing complete, reliable workflows within an open system architecture.

Further information can be found at www.printcity.de

Print Talk

Print Talk is a community of suppliers of e-business solutions and order management systems formed to define best practices and interfaces between products. Print Talk implementations support JDF and Commercial eXtensible Markup Language (cXML).

Further information can be found at www.printtalk.org

W3C

The World Wide Web Consortium (W3C) develops specifications, guidelines, software and tools to improve communication on the Web.

Further information can be found at www.w3c.org

WfMC

The Workflow Management Coalition (WfMC) is an international non-profit organization consisting of workflow suppliers, users, consultants, universities and research groups that aims to develop and maintain standards for software in the rapidly expanding market.

Further information can be found at www.wfmc.org

Companies

The following provides a list of the companies whose concepts and solutions are presented in this book. This of course only represents a small selection of the companies dealing with the issue of networking for print and media service providers.

Adobe

Adobe Systems Inc. is one of the initiators of the desktop publishing revolution and is also a pioneer in the next era of publishing, network publishing. Adobe is one of the initiators of the Job Definition Format.

Further information can be found at www.adobe.com

Agfa

Agfa is a solution provider for prepress, print production and publishing. Agfa is one of the initiators of the Job Definition Format and has also joined PrintCity.

Further information can be found at www.agfa.com

Creo

Creo is a solution provider for prepress with workflow solutions for both offset and digital printing. Creo is building up a network of partnerships for networking under the banner of Networked Graphic Production.

Further information can be found at www.creo.com

DiMS!

DiMS! organizing print develops, distributes and implements DiMS!, an upscale MIS for the printing and packaging industry.

Further information can be found at www.dims.net

EFI

Electronics for Imaging Inc. is developing and selling through diverse OEM Partners Server and controller to transform digital color copiers into networked color printers. It has diversified into new markets and distribution channels by acquiring prepress and MIS solutions for the commercial print market.

Further information can be found at www.efi.com

Global Graphics

Global Graphics Software Ltd. is a solution provider for the print media industry. Global Graphics has joined PrintCity.

Further information can be found at www.globalgraphics.com

Heidelberger Druckmaschinen

Heidelberger Druckmaschinen AG is a solution provider to the print media industry. The company's product portfolio comprises machines and software for print service providers' entire value adding process. The networking solutions of Heidelberger Druckmaschinen AG are offered under the name Prinect. Heidelberger Druckmaschinen is one of the initiators of the Job Definition Format.

Further information can be found at www.heidelberg.com

Hiflex offers an order management system with an integrated production planning system. Hiflex is a member of the NGP.

Hiflex

Further information can be found at www.hiflex.de

MAN Roland Druckmaschinen AG is a solution provider in the print media industry. The company's product portfolio comprises sheetfed and web offset presses. Networking solutions are offered under the name PeCom. MAN Roland founded PrintCity in order to market networking solutions together with other suppliers from the print media industry. MAN Roland is one of the initiators of the Job Definition Format.

MAN Roland

Further information can be found at www.man-roland.com

Optichrome Computer Systems Ltd. develops and sells the order management system Optimus 2020. OCSL has joined PrintCity.

OCSL

Olive Inc. is a Japanese company which develops and distributes Print Sapiens and Net Sapiens. It is a leading MIS supplier in the Japanese print media market.

Olive

Further information can be found at www.olv.co.jp

The software house Orgasoft GmbH develops and sells the OS ABSYS order management system.

Orgasoft

Further information can be found at www.orgasoft.org

A part of MAN Roland, ppi Media GmbH offers PrintNet, a JDF-based production planning system for newspaper printers and publishers.

ppi Media

Further information can be found at www.ppimedia.de

Prism, is an Australian Software House, which develops and distributes Prism WIN, QTMS, Enterprise 32, Prism online. It is a leading MIS manufacturer in the print media industry.

Prism

Further information can be found at www.prism-world.com

Industry standards

The following list contains some specifications and links to industry standards in the print media industry. This information can of course only reflect the situation at the time this book goes to press. The reader should therefore check whether later versions have become available in the meantime.

ICC ICC-Spec.1:1998-09, File Format for Color Profiles, version 3.5
Date: 1998
Produced by: International Color Consortium
http://www.color.org/ICC-1_1998-09.PDF

IFRAtrack IFRAtrack-Specification version 2.0
Date: June 1998
IFRA Special Report 6.21.2
Produced by: IFRA
http://www.ifra.com/

JDF JDF Specification version 1.1 revision A
Date: September 5, 2002
Produced by: CIP4
http://www.cip4.org/

PODi PPML Functional Specification version 1.1
Date: July 2002
Produced by: Print-on-Demand-Initiative
http://www.podi.org/

Print Talk Print Talk-Implementation version 1.0
Produced by: Print Talk Consortium
http://www.printtalk.org/

PJTF Portable Job Ticket Format
Date: April 2, 1999
Produced by: Adobe Systems Inc.
http://partners.adobe.com/asn/developer/PDFS/TN/5620.pdf

PPF Print Production Format version 3.0
Date: June 2, 1998
International Cooperation for Integration of Prepress, Press, and Postpress
http://www.cip4.org/documents/technical_info/cip3v3_0.pdf

XML XML specification version 1.0
Date: February 10, 1998
Produced by: World Wide Web Consortium (W3C)
http://www.w3.org/TR/REC-xml

XML Schema Part 0+1+2: Primer, Structures and Datatypes

XML Schema

Date: May 2, 2001

Produced by: World Wide Web Consortium (W3C) XML-Schema working group

http://www.w3.org/TR/xmlschema-0/(-1/;-2/)

Glossary

General IT terms

B2B Business-to-business (B2B) is a variant of e-commerce where both the purchaser and the vendor are companies. Increasingly, entire business processes are being fine tuned to B2B. The main benefits provided by such virtual platforms are the standardization of processes and data when working with external partners and the opportunity to tap into new procurement and sales markets.

CIM The term Computer Integrated Manufacturing (CIM) was coined at the start of the 1980s and associated with expectations of increasing automation culminating in the 'unmanned factory'. These expectations could not be fulfilled at the time, mainly due to a lack of standardized interfaces. Today, things have moved on considerably in many branches of industry. The realization that human skills are a valuable part of a company has led to the goal of the 'unmanned factory' becoming less and less popular.

E-commerce Electronic commerce (e-commerce) is used to describe transactions that take place over the Internet. E-commerce solutions offer the possibility of clarifying product information, price information and terms and conditions via the Internet. Payment can also be made over the Internet (credit card, online banking).

E-procurement Electronic procurement (e-procurement) is the use of Internet-based information and communications technologies to optimize procurement workflows. One e-procurement solution that is widely used in the print media industry is "Paperconnect", which print service providers, publishers and advertising agencies can use to procure fine paper and information about the paper market over the Internet.

Software applications

CRM

Customer Relationship Management (CRM) describes an integral approach to company management. CRM integrates and optimizes customer-related processes across departments in marketing, sales, customer service and research and development. In the print media industry, CRM modules are sold as part of the order management system. These modules are tasked with supporting field staff with data from the order management system when they are preparing quotations and providing customer care, and enabling orders to be transferred to the order management system without a change in media.

ERP

The basic idea behind Enterprise Resource Planning (ERP) is that of company-wide, and even cross-company, provision of data about production factors (production, material management, planning, logistics, HR, accounting) whose scope far exceeds that of traditional production planning and control (PPC) systems. The best-known examples of such software applications are SAP/R3 for large companies and mySAP for SMBs.

Machine data logging

Machine data logging embraces all the signals transferred from software applications and machines that allow conclusions to be made regarding resource usage or resource consumption. Machine data is logged automatically without operator intervention. It complements operating data logging.

MIS

Management Information Systems (MIS) are tasked with making specific information about company management available to managers. In the print media production sector, order management systems are sometimes also referred to as MIS.

Operating data logging

Operating data logging consists of manual inputs by staff in production with the goal of capturing the actual consumption of resources and events in production. To do this, staff use an operating data logging terminal or an operating data logging mask directly on the press control system or in the software application. Operating data logging complements machine data logging. The recorded data is made available to the order management system for actual costing and statistical evaluation.

Order management system

Order management systems are used to create, cost, plan and control orders from the quotation phase right up to completion. For historical reasons, these systems have also been known as costing systems and industry software and sometimes also as MIS (Management Information Systems) in the print media industry.

PPC Production planning and control (PPC) systems are software applications for planning materials, capacities and deadlines. Orders are planned using an electronic planning chart and master, order and production data is sent to the press control system together. Feedback from the machines enables the production process to be controlled in real time.

SCM The term Supply Chain Management (SCM) represents the optimization of the entire supply chain. It centers on the value added chain from subsuppliers via suppliers and producers to intermediary and end customers. In tandem with e-business solutions, technologies are used that lead directly to improvements in efficiency and cost reductions and enable close partnerships with customers and suppliers. Using the end customer as a starting point, SCM embraces cross-company management of material and information streams across business processes in the value adding network.

Data exchange formats

EDIFACT

The Electronic Data Interchange For Administration, Commerce and Transport (EDIFACT) format has been used successfully for years to exchange commercial documents and business correspondence electronically. Converters are used to transform incoming EDIFACT messages into the format used by the software application.

JDF

The Job Definition Format (JDF) is a manufacturer-independent data exchange format for defining and executing jobs that is intended to cover the entire process chain from the generation of quotations to final delivery. JDF combines the existing formats Print Production Format (PPF), Portable Job Ticket Format (PJTF) from Adobe and IFRAtrack with as yet non-standardized order data to form a new, Internet-compatible standard. JDF is based entirely on the XML standard. JDF is maintained and developed by the CIP4 Consortium (International Cooperation for the Integration of Processes in Prepress, Press and Postpress).

JDF templates

JDF templates are reusable templates for products, product families, production types and production steps for quickly and effectively constructing JDF from existing information.

JMF

The Job Messaging Format (JMF) is an open-standard format that, in an extension of JDF, enables dynamic communication to take place even while processes are running. JMF is also based on the XML standard and is maintained and developed by the CIP4 Consortium (International Cooperation for the Integration of Processes in Prepress, Press and Postpress).

PDF

The Portable Document Format (PDF) from Adobe, a cross-platform data output format, is a recognized standard in the print media industry for content data that has also become widely adopted in the IT sector in general.

PJTF

The Portable Job Ticket Format (PJTF) was developed by Adobe for attaching control commands for output and finishing devices to PDF files. PJTF files contain information for processing pages, information on output parameters and even CIP3 information and administrative data. PJTF files can be embedded in the PDF files and enable output settings to be made without having to change parameters in the PDF file. The functions of PJTF are included in JDF.

PPF

The Print Production Format (PPF) is a forerunner of JDF that was devised by the CIP3 (International Cooperation for Integration of

Prepress, Press and Postpress). PPF files are used in prepress to describe numerous parameters for a print product. These parameters can be used later to preset presses and finishing machines. With JDF, a separate CIP3 interface is no longer required, since all the functions are contained in the JDF.

XML eXtensible Markup Language (XML) is a meta-language standardized by the W3C and used for defining document types. This extensible meta-language is platform-, application-, language- and domain-independent.

Business management terms

Benchmarking is a tool for strategically controlling value adding processes, management practices, products and services between companies or business units with the aim of discovering shortcomings in performance. *Benchmark*

Controlling is a planning, checking, management and coordination tool used by management to better monitor and manage complex processes, in particular the effectiveness of operational measures. *Controlling*

The contribution margin is the difference between sales proceeds and variable costs. The contribution margin serves to cover fixed costs. The part that exceeds the fixed costs is profit. *Contribution margin*

Cost centers are performance areas of a company defined for analytical purposes, e.g. presses, groups of presses, workstations. *Cost centers*

Cost units are a company's individual products (services sold and services rendered within the company), the proceeds from which are used to cover costs. *Cost units*

Direct costs are types of costs that can be directly allocated to the cost centers or cost units. *Direct costs*

In a full cost system, all costs are allocated to the relevant allocation bases (e.g. cost centers, cost units). *Full cost system*

Imputed costs are costs to be accounted for in the profit and loss account statement that do not correspond directly to any expenditure, and therefore must be treated so as not to affect the operating result. *Imputed costs*

Indirect costs cannot be directly allocated to any allocation base and are converted in the cost analysis using a compensation key (indirect cost allocation). *Indirect costs*

The net monetary value represents the discounted, accumulated value created by an investment. *Net monetary value*

The payback period (payoff period) designates the period of time it takes for all outlays to be covered by returning income. *Payback period*

Payment-related refers to expenditure or earnings affecting payment. *Payment-related*

The percentage return on sales is the relationship between profits and sales expressed as a percentage, and therefore represents an important criterion when monitoring success, comparing operations and deciding on investments. *Percentage return on sales*

Return on Investment The Return on Investment (ROI) is a key figure for determining the profitability of investments and is defined as the product of the percentage return on sales (ratio of profits to sales) and capital turnover (ratio of sales to capital invested).

Workcenter calculation The workcenter calculation determines the costs related to a given period of time per cost center, e.g. per press, workstation etc.

Subject Index

A

Acceptance, 68
Acceptance test, 73
Accounting, 56
Activities, 68, 69, 71
Actual costing, 9, 11, 16, 17, 55, 56, 66, 71, 74
Adobe, 8, 21, 31
Agfa, 21
Agreement, 16
Annual accounts, 17
Application of dampening solution, 9
Area coverages, 13
ASCII file, 34
AuditPool element, 28
AuditPool Elemente, 29
AuditPool elements, 28, 29

B

Banks, 17
Benchmarks, 68, 75
Binding, 27
Budgets, 51

C

Calibrated, 17
Calibration, 18
Calibration curves, 8
Capacity planning, 16
Capacity utilization rate, 75
Certification, 8, 9
Change of media, 4
Changes, 12, 13, 15, 29
CIP 3 Consortium, 14, 31
CIP 4 Consortium, 22
CIP3, 14
CIP3 Consortium, 8
CIP4, 22, 23, 31, 32
Color, 6
Color control bar, 18
Color management, 69
Color quality, 18
Color workflow, 17, 18, 57
Company management, 55
Compatibility, 30
Competitiveness, 75
Complaints, 18
Consultant, 68
Consumables, 16, 56
Content data, 5, 6, 11
Contribution margin, 70, 75
Control consoles, 13
Control data, 5, 8
Controlling, 74, 75
Conversion ratio, 75
Corporate design, 51
Cost center, 69, 75, 76
Cost center statistics, 75
Cost drivers, 69
Cost pools, 69
Costs, 65, 66, 69, 75, 76
Costs incurred due to errors, 52
Creo, 18
Customer analysis, 75
Customer loyalty, 11, 12, 50, 75
Customer master data, 26
Cutters, 54
Cutting, 27
Cutting marks, 54

D

Daily dockets, 9, 55, 56
Data format, 6, 21, 32
Data streams, 10
Data types, 3, 5
Database, 6, 24, 26
Degree of utilization, 13
Delivery, 10, 16
Delivery dates, 15
Digital printing, 11, 33, 50, 85
Division of labor, 10
Document, 6
Duplicated work, 76

E

E-business, 11, 12, 32, 49, 50, 85
Electronic job ticket, 13
Encryption procedures, 34
ERP, 12
ERP system, 33, 50
Error costs, 76
Ethernet, 25
Euro, 62
Evaluation, 9
Exchanging blankets, 18

F

Field staff, 55
Financial accounting system, 17
Finishing, 13, 16, 27
Folders, 54
Folding, 27
Folding type catalogs, 8
Form production, 69
Functional specification, 67, 68, 70, 73

G

Guidelines, 18

H

Heidelberger Druckmaschinen, 18, 21
Hot folders, 30
Hourly rate, 76
http
– //www.cip4.org, 8

I

IFRAtrack, 21
Image Control, 62
Image databases, 11
Incorrect entries, 55
Indirect cost allocation rates, 69, 70
Indirect costs, 69, 70
Industry-specific software, 7
Ink, 18
Ink changes, 15
Ink zone values, 13
Installation, 70
Interfaces, 17, 24, 25, 66, 71, 85
Internet, 6, 7, 11, 12, 25, 32, 34, 51
Internet portals, 50
Intranet, 34
Inventory, 67
Investment, 14, 76
Investment analyses, 59
Investment analysis, 58, 60
Investments, 4
Invoice, 50
Invoicing, 52
IRD, 49, 52, 53, 55
ISO, 9
IT departments, 12

J

JDF, 3, 8, 21–34, 66, 70, 72
JDF interfaces, 66
JDF template, 27
JDF templates, 27
JMF, 29
Job costs, 70
Job data, 5, 6, 13
Job Definition Format, 33
Job planning, 11
Job planning and control, 11
Job preparation, 4, 7, 11–13, 27, 52, 53, 61, 62, 69, 71, 72
Job preparation stations, 13, 14
Job processing, 33

Job throughput times, 52
Job ticket, 6, 13
Job tickets, 12, 52, 66
Jobs, 4

L

Level 3 RIP, 33
Libraries, 23
Life cycle, 70
Logging, 21
Logistics, 4

M

Machine data, 16, 56
Machine data logging, 66
Machine downtimes, 76
Machine presetting, 11, 13
Machine presettings, 53, 54
Maintenance, 76
Maintenance operations, 15
MAN Roland, 18, 21
Margins, 4
Master data, 5, 6
Messaging element, 29
Meta-language, 24, 94

N

Network, 25, 34, 71
Networked Graphic Production, 18
Networking, 10, 12–14, 17–19
Networking architecture, 23, 66, 70
Networking routes, 3, 9, 11, 18, 73
Networks, 25
NodeInfo, 27
NodeInfo element, 27
Non-productive times, 76

O

Operating and machine data, 5, 9
Operating data, 9, 16, 56
Operating data logging, 11, 16, 55, 56
Operating data logging terminals, 17, 55
Operational organization, 71
Operational organizations, 71
Operations, 69
Optimization, 74
Order confirmation, 50
Order management, 14, 55, 66, 71, 72
Order management system, 12, 16, 65
Order planning, 15
Order planning and control, 15
Order preparation, 14
Order processing, 4, 11
Order structure, 11
Orders, 33
Organizational structures, 71
OSI 7-layer model, 24
Output devices, 17
Overview of results, 75

P

Paper, 6, 18
Parameters, 25, 33
Payback period, 49, 51, 58–60
Payment streams, 32
Payment transactions, 17
Payment-related, 60
Payroll, 17
PDF, 22, 32, 33
PDF/X3, 32
PeCom, 18
Periodicals, 15
PJTF, 7, 8, 21, 31, 32
Planners, 16
Planning, 15
Plates, 15, 16
Platesetters, 4
Postpress, 4
Postscript, 8
Potential improvements, 67, 68
Powder length, 9
PPF, 13, 14, 21, 31
PPML/VDX, 33
Preflight, 69
Preflight checking, 33
Prepress, 4, 7, 11, 13, 14, 16, 18, 32
Prepress workflow, 65, 69
Presettings, 14
Press, 13, 15, 18
Press data, 9

Presses, 4, 9
Previews, 13
Principle of average, 69
Prinect, 18
Print media industry, 5, 21, 33, 85
Print Production Format, 14
Print Talk, 32
PrintCity, 19
Printing units, 15
Private section, 30
Private sections, 30
Process costs, 4, 11, 12, 18, 50
Process costs analysis, 67, 68
Process description, 27
Process deviations, 18
Process disruptions, 75, 76
Process fluctuations, 76
Process groups, 27
Process resources, 27
Process times, 27
Product description, 26
Product group analysis, 75
Product groups, 70, 75
Product mix, 68, 70
Product node
- Node, 28
Product nodes, 27
- Nodes, 27
- Nodes, 25, 27
Product structure, 26, 27
Production, 4, 10, 16, 19
Production control, 9, 21
Production data, 5, 7, 8, 12
Production guidelines, 76
Production hours, 75
Production plan, 15
Production planning, 15
Production planning and control, 54, 71, 72
Production planning system, 54
Production process, 12, 18
Production processes, 14, 27
Production reliability, 12
Production resources, 7, 15
Productive and non-productive times, 75
Profiled, 17
Profit, 76
Profits, 75
Project, 12
Project management, 65
Proprietary formats, 7, 14
Protocols, 24, 25
Provider, 12
Publisher, 6

Q

Quality, 9, 54, 75
Quality checks, 76
Quality data, 5, 9
Quality level, 13
Quotation, 67

R

Random checks, 56
Real time, 16
Real time operation, 73
Real-time operation, 73, 74
Reference values, 18
Remote proofing, 8
Reorganization, 71
Requirements analysis, 67, 68
Resource, 27
ResourceLink elements, 27
Resources, 27, 71
Return on Investment, 59
Return on sales, 75
ROI, 61

S

Saddlestitcher marks, 13
Saddlestitchers, 54
SAP, 33, 34
Savings, 71
Scanner, 17
Schedule, 15
Security, 34
Security guidelines, 12
Servers, 71
Service, 11, 12, 70, 71, 75
Setup times, 13, 53, 54, 62, 63
Seybold Conference, 21
Sheet, 18

Simulation, 29
Simulations, 54
Specification, 68
Spectrophotometric measuring devices, 18
Staff, 9, 10, 13, 66–68, 72
Standard, 22, 23, 30, 32, 33
Standard print products, 11
Standardization, 17
Statistics, 74, 75
Status, 9
Status messages, 55
Storage solutions, 71
Supplier, 70
Suppliers, 10, 16, 70
System supplying KPIs, 74

T

Target/actual comparisons, 66
TCP/IP, 25, 30
Tolerance range, 18
Tolerance zone, 9
Transaction costs, 11
Transparency, 13
Trapping, 69
Tree structure, 27, 28

U

Used-capacity principle, 69
User interface, 12
Utilization, 9, 15

V

Verifying success, 74

W

W3C, 24, 94
Waste, 13, 53, 57
Waste paper, 18
Working time, 14
Works council, 16, 65

X

XML, 24, 25, 30, 32–34, 94